THE MAKING OF YOU

KATHARINA VESTRE

THE MAKING OF YOU

The Incredible Journey from Cell to Human

Translated from Norwegian by
MATT BAGGULEY

Illustrations by
LINNEA VESTRE

GREYSTONE BOOKS
Vancouver/Berkeley

First published in Norwegian in 2018 as *Det Første Mysteriet* by
H. Aschehoug & Co., Norway
Published in agreement with Oslo Literary Agency

19 20 21 22 23 5 4 3 2 1

Greystone Books Ltd.
greystonebooks.com

Cataloguing data available from Library and Archives Canada
ISBN 978-1-77164-492-1 (cloth)
ISBN 978-1-77164-493-8 (epub)

Jacket and text design by Belle Wuthrich
Jacket illustration by Linnea Vestre
Proofreading by Alison Strobel

Printed and bound in Canada on ancient-forest-friendly paper by Friesens

Greystone Books gratefully acknowledges the Musqueam, Squamish and Tsleil-Waututh peoples on whose land our office is located.

Greystone Books thanks the Canada Council for the Arts, the British Columbia Arts Council, the Province of British Columbia through the Book Publishing Tax Credit and the Government of Canada for supporting our publishing activities.

Canada

BRITISH COLUMBIA
BRITISH COLUMBIA ARTS COUNCIL
An agency of the Province of British Columbia

Canada Council for the Arts
Conseil des arts du Canada

CONTENTS

PREFACE

WHEN I WAS six years old I collected hotel soaps, played with Barbie dolls and wore flashing sneakers. My taste in movies was exceptionally unoriginal and can be summed up as 'anything with princesses'. But my favourite book? *Pregnancy and Birth: A Practical Handbook for All Future Parents*. My sister and I would take it down from the bookshelf, skim past all the dietary advice and stop when we reached page 70: *The foetus as it grows*. Fascinated, we would follow the illustrations of this tiny creature as it increased in size, thinking of our own brother-to-be curled up inside our mother's tummy. We learned how he was transforming from a strange, primitive little animal with a tail into a chubby baby with arms and legs, confined in a space barely large enough to accommodate him. How was this possible?

Seventeen years passed before I returned to this question. I was completing a bachelor's degree in biochemistry at the University of Oslo, and sitting up late in the library one night, reading about cell biology. In my textbook was a series of images showing how a hand is formed in the uterus. At first it resembles a duck's foot, and then the fingers gradually appear. In the caption I read that this transformation was due to mass cell suicide. Many years ago, all the cells that linked my fingers together died, on command from their neighbours, and left me with the hands I'm writing with now.

This detail, I realised, was not included on page 70: *The foetus as it grows*. The pictures I'd seen as a six-year-old told only a small part of the story. How does this tiny creation *actually* come about? What happens in the cells, and in the DNA molecules? How does a hand know that it's going to be a hand and not a foot or an ear, for example?

In search of answers, I began digging through syllabus books and research articles. It wasn't long before I became completely immersed. Prior to the summer vacation in 2015, I borrowed three huge embryology books from the library at Oslo University Hospital and took them with me on holiday to Italy. My internet search history filled up with egg cells and foetuses.

Google drew its own conclusions and began showing adverts for baby creams (I don't like to think what their algorithms made of my searches for fruit flies, fish kidneys and the gender development of sea worms).

The result of all this was the book you now hold in your hands. It is a story about distant relatives, unknown twins, dangerous placentas and strange insects. And I can say right now – without giving too much away – that it is all about *you*. Let me tell you about the beginning of your life.

Before we begin: a few words about time and size

WHILE WORKING ON this book, I discovered that trying to state the age of a foetus is fraught with difficulty. There are various chronological calculations involved, and it's not unusual for them to get mixed up. Doctors and midwives commonly use the *week of pregnancy*, which is calculated from the last menstruation. However, conception usually occurs about two weeks after this, so it's not until a woman begins her third 'week of pregnancy' that she's actually pregnant. In other words, the foetus is two weeks younger than the week of pregnancy: at the end of the twelfth week of pregnancy, the foetus is ten weeks old; at the end of the fourteenth week it's twelve weeks old, and so on.

I've chosen to use the conception as my starting point, so that all the time references I give reflect the real age of the foetus. Next, what is meant by a 'month'? I have counted each month as a four-week period rather than a calendar month. Thus the first month comprises weeks one to four, the second weeks five to eight and so on.

When I state the length of a foetus I mean the measurement from its crown to its rump. (You will sometimes hear this referred to as CRL, crown-rump length.) This measurement is preferred because the legs of a foetus are often bent upwards, making it difficult to establish its length from head to toe.

Finally, please keep in mind that all time and size references are based on average values, and that every foetus develops at a slightly different rate. So, with that said, I think we're ready to begin.

THE RACE

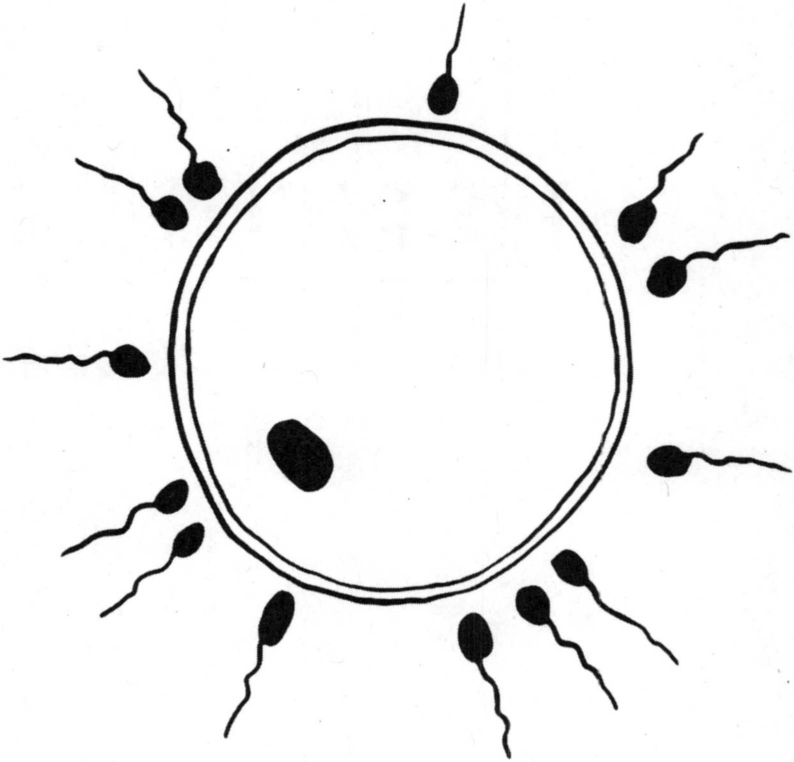

Fond du Lac Public Library
32 Sheboygan St.
Fond du Lac, WI 54935
(920)929-7080

Items that you checked out

Title: Radical compassion : learning to love yourself and your world with the practice of RAIN
ID: 39044005262152
Due: Friday, March 27, 2020

Title: The making of you : the incredible journey from cell to human
ID: 39044005057503
Due: Friday, March 27, 2020

Total items: 2
3/6/2020 4:49 PM
Checked out: 11
Hold requests: 2
Ready for pickup: 0

Check us out on the web at:
www.FDLPL.org

IN THE HOURS preceding conception, a race begins that is almost impossible to win. A sperm cell starts out on an intense swimming trip. It looks like a little tadpole, battling wildly upstream against the current and in unknown terrain. It has several hundred million competitors. And it must swim a distance more than one thousand times its own body length. The rules are simple: reach the finishing line first, or die.

The landscape around the sperm is confusing and inhospitable, an overgrown forest full of chaotic thickets and blind alleys. It risks being either swallowed up by immune cells or destroyed by acid on the way. It could also end up trapped in one of the deep crevices in the cervical wall. Before long, such hazards have eliminated most of the field, but our competitor is luckier: the woman's muscle contractions help to push it upwards and it manages to enter the uterus.

It is still a long way from victory. To have any chance of winning, it must first choose where to go next: right or left? The uterus is connected to two narrow channels – the fallopian tubes – and the finish line is at the end of one of them. The walls of each tube are lined with hairs that sweep fluid down to the uterus, but the sperm cell refuses to give up. It struggles on against the flow. Somewhere up there, hidden among the deep valleys and high peaks of the mucous membrane, a round egg is about to meet its champion.

The egg cell has waited a long time for this moment. When the woman was a tiny foetus herself she'd already made the forerunners of her egg cells. Later on she began transforming them into mature egg cells. The cell now floating down her fallopian tube was one of the lucky ones. Each month, several egg cells start maturing inside every fertile female, but only one of them gets the chance to escape. The others face certain death.

To create a mature egg cell, the forerunners divide so that the chromosome pairs containing the genes from the new foetus's grandmother and grandfather are separated. At the end, each mature egg cell has half a set of chromosomes, some from Grandma and others from Grandpa, ready to find itself a new partner. All the while, the egg cell has been packing itself with nutrients, blowing up like a giant compared with the other cells in the body. It's actually possible to see the egg cell without a microscope: it has a diameter of around a tenth of a millimetre.

The sperm cell could not be more different. Swimming frantically with its wriggling tail, there's barely any room for nutrients because its entire head is packed with the father's DNA. Among the many millions of sperm cells, only one of them carries half of your specific genes; the chances of two sperms being identical are vanishingly small. Had another of your father's sperm swum just a little faster, you would never have existed as you are now.

When your parents' sperm and egg cells were formed, the chromosomes from your grandparents sat right next to each other; and before the chromosome pairs were separated from each other for ever, they managed to exchange small pieces of DNA. So a chromosome that was originally from a grandmother can carry some genes from a grandfather when it ends up in the sperm cell. The possible combinations are endless, and so we have to be sure we cheer on the right sperm.

Returning to our race commentary, our frenetic little tadpole is made for what it's doing right now. It may be blind and deaf, but that doesn't stop it making its way through a landscape it's never even been close to before. Among other things, it can sense minute changes in temperature. Because its target is slightly warmer than its surroundings, the sperm can tell when it is getting close. In addition, the sperm is equipped with a basic sense of smell. Just as the cells in your nose do, sperm cells contain molecules called odorant receptors. Each odorant receptor is programmed to recognise a particular molecule. When air flows into your nose, the fragrance molecules become attached to different odorant receptors and create an electrical signal that is sent to the brain. In the case of sperm cells, the odorant receptors catch molecules streaming from the egg, confirming that it is on the right path.

At the finish line there are relatively few competitors left swimming, and the egg's attractor chemicals make them travel faster than ever. Soon the egg is completely surrounded by minute tadpoles. Their tails wriggle furiously as they drive forward into the jelly-like membrane protecting their goal. From their heads they spray chemical weapons, enzymes that break down the membrane and allow them to burrow even deeper.

But only one of them is fast enough. The winner discards its tail, melts into the egg and releases its valuable cargo: twenty-three of the father's chromosomes. At the same instant, the egg cell releases substances that create a hard capsule around it so that no more sperm can enter. There's no time to lose: if two sperm cells were to penetrate the egg at the same time, the result would be a cell of sixty-nine chromosomes instead of forty-six. Although the egg cell does its best to avoid this, it isn't always successful. When a group of researchers studied artificially fertilised eggs, they found that 10 per cent of them had been fertilised by more than one sperm cell. Eggs like this have no chance of developing normally, and, as we'll see later on, they are handed a death sentence.

But for now you can relax – this time there was only one winner. The chromosomes from your mother and father are now united and your very first cell has been created. The race is over. The making of you can begin.

THE HIDDEN UNIVERSE

BEFORE MICROSCOPES CAME along, most of what happened at the very beginning of human life was hidden from us. With the naked eye it is almost impossible to see the minute details gradually emerging. Even elephants, which rise four metres above the ground, start off microscopic. It doesn't help either that we are concealed behind skin, muscles and blood vessels.

More than 2,300 years ago Aristotle wondered how new creatures might come about. In search of answers, he opened fertilised chicken eggs at different stages of gestation. In a three-day-old egg he observed a little red heart beating within the yellow yolk. When he cracked open a shell after a week, he found a tiny creature with large eyes. Of course, the later he broke the egg, the more the embryo resembled a chicken. Surely, he thought, it was the same with people too. Aristotle surmised that a man's sperm somehow instructed the woman's blood to create a human in her stomach.

That said, Aristotle also believed that living creatures could arise in very different ways. Insects could be created from the dew on leaves, moths from wool and oysters from slimy mud. Almost two thousand years later, these ideas were still popular. In the seventeenth century the Flemish chemist Jan Baptist van Helmont came up with some highly creative and entertaining recipes for the world's various life forms. For example, let's say you fancy growing some mice. The recipe for this is simple: place a dirty, slightly sweaty shirt into a container full of wheat grain. Wait twenty-one days and – *voilà!* – your wheat has been transformed into sniffing, twitching mice.

There's no reason to doubt that van Helmont's recipe worked. Neither was he alone in providing striking examples of how animals could appear, quite spontaneously, if the conditions were right. Wet mud along the riverbanks could magically transform itself into frogs, rubbish into rats, and just imagine all the white larvae appearing from nowhere on rotten meat. And it was hard to imagine how oysters could possibly mate – surely they just somehow sprang into existence.

At the end of the 1600s a new idea emerged: perhaps every creature, be it a frog or a human, arose from a miniature version of itself. When God created the first humans in all their perfection, he also created all future generations. These miniature humans were nested inside each other, layer upon layer, like Russian dolls. Later they would simply germinate and grow in their mother's womb until birth. When microscopes first arrived, biologists grew even more confident that they would discover these scaled-down creatures existing somewhere. Just imagine the riches of detail that lay hidden from the eye! There seemed to be no limit to what could be found if only microscopes might improve just a little more.

One of the most talented microscope makers of the time was a Dutch merchant named Anton van Leeuwenhoek. There was little in his background to suggest that he would end up a scientist: he had no university education and no wealth. His original motivation was simply to investigate the quality of the textiles he sold. Nevertheless, one day Leeuwenhoek became curious and placed a drop of water under his lens. What he saw changed his life for ever. Each transparent droplet was teeming with mysterious creatures of every possible shape. Leeuwenhoek named them *animalcules* (tiny animals), and soon began to investigate everything he came across: the water he drank, the puddles he stepped in – even the deposits he found between his teeth.

Everywhere he looked, he found tiny animals. Forget exotic islands, forget about space, Leeuwenhoek could peer into a secret universe – barely explored – right before the tip of his nose.

Rumours of Leeuwenhoek's impressive microscopes spread fast. One day he was visited by a medical student who brought with him a sperm sample taken from a sick patient. Leeuwenhoek had for some time declined to study sperm; as a religious man, he feared that it would be considered profane. On the other hand, this *was* clearly a medical case … He resolved to take a look. The sample he examined wasn't much larger than a grain of sand. And yet, under the lens he could see more than a thousand minuscule creatures. They had round heads and long, transparent tails – like tiny tadpoles. Had they come about because of the man's disease? Had the sample been stored for too long, perhaps?

Like any good scientist, Leeuwenhoek realised that he had to compare his observations with a sample taken from a healthy subject. In 1677 he reported his findings in a letter to the president of the Royal Society of London – one of the world's leading research institutes –

in which he gave a detailed description of the animals he'd observed in the healthy sample, and wrote that it was examined 'immediately after ejaculation, before six pulse strokes had passed'. Afterwards, he was keen to emphasise that the sample was, of course, not obtained in any sinful way, but 'made available' to him 'quite naturally, through marital activity'. (It couldn't have been easy to be his wife.) At the end of the letter, Leeuwenhoek requested strongly that the president keep their correspondence to himself, should he feel that the observations risked causing disgust among the scholars. A scandal was the last thing he wanted.

Leeuwenhoek was convinced that semen played a decisive role in the beginning of life. This was no clear, empty fluid – it was packed with swarms of microscopic creatures! Could this be the place that miniature humans existed? Surely he needed only a good enough microscope to reveal it. For years Leeuwenhoek worked persistently, but despite continually improving his lenses, he found nothing. He even attempted to remove the membrane around the sperm cell's head with a tiny brush in the hope of finding something hidden inside. Despite the absence of proof, Leeuwenhoek was certain that the sperm cell contained a great secret – albeit one so small that humankind would never be able to see it.

THE RECIPE FOR A HUMAN

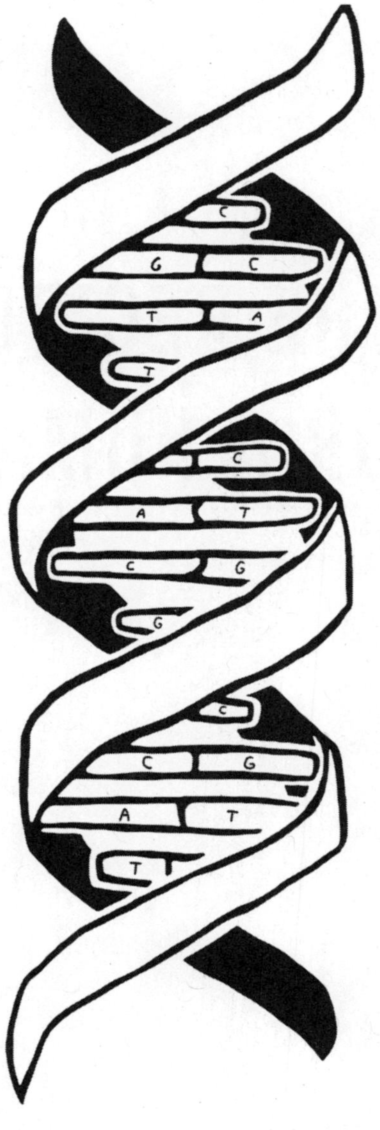
C
G
C
T
A
T
C
A
T
C
G
G
C
C
G
A
T
T

THE FIRST FEW hours. The race is over, and your very first cell floats calmly down the fallopian tube. So much has already been decided. Even though the cell is tinier than the full stop at the end of this sentence, it is large enough to contain all the instructions needed to build you, from the organs that will keep you alive to your eye colour and the shape of your nose.

Leeuwenhoek never managed to discover it, but the cell's great secret was not a miniature person. It was a molecule. And the story of this molecule begins with pus.

In 1869 a Swiss chemist named Friedrich Miescher contacted a surgical clinic near his laboratory. 'Would you let me have some used bandages from your patients?' he asked. 'Preferably covered in as much pus as possible.'

Miescher was looking for white blood cells, something there were masses of in the whitish-yellow gloop that oozes from sores. White blood cells are the remnants of a battlefield – they work for the immune defences – and many of them give their lives in the war against bacteria that is fought in wounds. Miescher collected the pus, filtered out the cells and conducted a thorough chemical analysis to examine the types of proteins they contained. During one experiment he noticed a sticky, milk-white substance that separated from the rest of the mixture when he added acid. Upon further examination, Miescher realised that it couldn't be a protein. He called the new substance *nuclein*, since it sat at the cell's core, the nucleus. Miescher found that there was an unusually large amount of nuclein in sperm cells, and realised that this substance must play a crucial role at the start of life.

Back then, genetics was still a mystery, driven by invisible forces that no one understood. The idea that hereditary material was actually a specific molecule – one that could be weighed and measured – was inconceivable for most people. But Miescher himself was an avid proponent of this theory. He suggested that the information contained in hereditary material was stored as a chemical code. It took a very long time for anyone to appreciate just how close Miescher was to the truth.

In the years that followed, many scientists studied the mysterious nuclein substance more closely. They discovered that it contained a type of sugar called deoxyribose, and that it was acidic. It was accordingly given the more precise name deoxyribonucleic acid, which was shortened to DNA. For many years DNA was looked upon as nothing more than a supporting material, holding things in place at the cell's core. The researchers realised that the genes must be in the chromosomes, but even then, the DNA molecule still failed to get the recognition it deserved.

Chromosomes consist of both DNA and proteins, and it seemed far more likely to the researchers that it was the proteins that controlled the genetics. From a chemical perspective, the proteins were more interesting – they existed in an infinite number of forms, with many different chemical and physical properties. DNA, on the other hand, looked to be the same everywhere. But when scientists at the Rockefeller University in New York experimented with bacteria in the 1940s, the results showed what everyone had thought was impossible: genes were made of DNA. How could this simple substance create the multitude of characteristics found in nature? White and pink blossom, curled and straight fur, pointed and rounded noses – were all these contained in the same molecule?

Only when James Watson and Francis Crick presented their model of the DNA molecule's structure in 1953 did the pieces fall into place. DNA was no aimless, disorganised lump. It was a chemical code. The molecule consists of long chains of four different bases: adenine, thymine, cytosine and guanine – A, T, C and G – which are connected to sugar and phosphate. Two chains connect themselves to a spiral staircase of base pairs. The sugar and phosphate provide supporting stringers: the base pairs form the steps going up. When the bases connect like this, they follow strict rules: an A will always connect to a T while a C will always connect to a G. Which means you will know exactly how one side of the ladder will look if you know the other. The cell can open the molecule in the middle and read these bases, letter by letter, like a book. By attaching matching letters on each side, it can create two identical copies of the recipe that can be passed on. Cell by cell, generation after generation.

A, T, C, G. These four letters are all that is needed. They can code eyes, fingernails, dimples, oak trees, jellyfish, sea grass, elephants, butterflies. Chemically, there's very little difference between the recipe for a human and that for an oak tree: the building blocks are exactly the same. It all depends on the order in which they are placed.

When that very first cell of yours floats down the fallopian tube, there are forty-six chromosomes sitting safely at its core. Twenty-three of them come from your mother and twenty-three from your father. Each one consists of a long DNA strand, tightly coiled around beads of proteins; which altogether adds up to over two metres of DNA in your one cell. The recipe was fixed when the sperm and the egg cell merged. Now it is time to use it.

THE INVASION

THE FIRST WEEK

DAY 3

IT IS ONE day after your conception, and the small hairs in the fallopian tube are nudging the tiny round cell down the canal. Slowly. Carefully. On the outside, everything looks perfectly calm. Deep within the cell's interior, however, a sophisticated mechanism is working tirelessly, creating precise copies of your DNA molecules. Soon, each chromosome becomes an X-shape, formed by two identical DNA molecules, attached at the centre. The chromosomes assemble at the centre of the round cell, row upon row. At the same time, the cell spins a web of protein from either side, and long thin strands reach towards the centre taking hold of the chromosomes. The cell then stretches and becomes elongated, while the strands pull the DNA copies towards each pole. It's beautiful to observe under the microscope, like a tiny firework. Finally, after about a day, it pinches in the middle and splits into two cells.

And it continues like that. The cells copy and divide, copy and divide. Some creatures, like bacteria or amoebas, are happy to consist of only one cell. They're still able to eat, move and multiply, and they have never wanted anything more. The male roundworm *Caenorhabditis elegans* consists of exactly 1,031 cells; we know this because biologists have taken the trouble to count them. But you? You consist of roughly thirty-seven thousand billion cells. 'Roughly', because there's such a crazy number of them that no one would ever sit down and count them one by one. Instead, scientists have calculated roughly how many there should be, knowing what we know about the body and the cells it is composed of. Even this is no simple matter. Cells come in myriad sizes, and how close they are to each other varies significantly. So, thirty-seven thousand billion give or take a few billion. And, incredibly enough, all these cells manage to cooperate. The amoeba might whirl around unrestrained, but your cells will form a tight-knit community.

First, though, they have to increase their numbers. During the first few days, the cells divide in a big hurry. They don't even bother growing, they just keep dividing into ever smaller units. Two cells become four. Four become eight. Soon you're a tiny cluster of sixteen round and totally identical cells. Under a microscope, they look a bit like a raspberry.

For about five days this little raspberry drifts calmly along the fallopian tube, until at last its food reserves are spent. For several days the cells have made do with leftovers from the egg. Now they are screaming out for a new form of nutrition. The time has come. The outermost cells take charge.

They begin pumping fluid from the fallopian tube into the centre of the cell cluster. This marks their first division of labour: from now on, the cells are no longer identical. The raspberry transforms into a vesicle, a sac with a fluid-filled cavity, which soon leaves the fallopian tube and enters the womb. It floats in there for a while as the cells continue to divide. And then, about a week after conception, a brutal invasion begins.

In the womb, your mother has prepared a thick, sponge-like membrane that the vesicle can stick to. Soon after, the vesicle expels a substance that causes the membrane to disintegrate so that it can burrow deeper. At this point, the whole scene resembles a gory horror film. Blood vessels burst. Cells die in their masses. Your ravenous cells feed on the mucosa that seeps out of the womb lining and sprout small roots that attach to your mother's blood vessels. That is how the placenta begins.

When you are born, the placenta is a slimy, blue-red slab weighing about half a kilo. It is ejected from the mother's womb just moments after a wriggling, screaming baby, so perhaps it's not so strange that it doesn't grab our attention there and then. Chubby arms and tiny little fingers are, after all, more instantly appealing. But in the past, the placenta was highly valued in many cultures. In ancient Egypt it was treated with the utmost care and then mummified. In Korea the placenta of a newborn prince or princess was placed in an ornate jar and buried. Even today, some people think it is a good idea to eat the placenta (Google, ever helpful, recently suggested I search for 'placenta smoothies'). There are companies that, for a fee, will freeze-dry a placenta and turn it into pills for the new mother to take. Either way, the placenta deserves a little gratitude. For almost nine months this peculiar organ works tirelessly for you. Without it, you would not be alive.

Those tiny roots, however, are just the beginning. Soon the invading cells will paralyse your mother's blood vessels and rebuild them according to their own needs. Her blood will leak out and fill up spaces in the placenta, and your veins will branch out to reach them, by winding their way through the umbilical cord. Your blood is never actually in direct contact with your mother's, but a great many substances can pass through the thin walls separating you. Because of this, you get all the oxygen and nutrition you need from your mother, and all your waste material is sent back to her in return. But it doesn't stop there. You also exchange hormones, and because of this, you and your mother can affect each other's bodies. The placenta quickly begins to produce a cocktail of hormones, which keeps your mother's blood vessels open and, among other things, makes her eat more. Furthermore, these hormones make sure that her body prepares itself for pregnancy and breastfeeding.

One of the hormones the placenta cells rapidly begin to make is called hCG. Regular pregnancy tests check a woman's urine for this hormone. Nowadays it's simple to take a test at home, but it wasn't quite so easy in the past. Back then, the doctor would need to sacrifice a mouse to get to the answer. Mice react in a specific way to the hCG hormone, and early pregnancy tests would therefore involve a doctor injecting a mouse with some of the woman's urine. A few days later, the doctor would kill the mouse and examine whether its ovaries had changed. The method was developed further in the late 1920s, and a few years later rabbits, which proved to be quicker and easier to work with, replaced the mice. The expression 'The rabbit died' became synonymous with 'I'm pregnant', although the animal bit the dust no matter what the test result was. More effective pregnancy tests – ones that did not involve animals – didn't become available until the 1960s.

Women have developed strict vetting systems so that not just anyone can set up home in their bodies. Only if the vesicle can confirm itself by sending the correct signal will it be allowed to stay put. It's possible that only about a third of the vesicles make it past the checkpoint, perhaps even fewer. Many pregnancies end without the mother ever realising they'd started. For example, an egg fertilised by more than one sperm cell will never pass this point. The extra chromosomes disrupt the neat web the cell normally weaves as it divides. Some of the cells end up with too few chromosomes, others with too many. If the cells aren't already in the process of dying, they're guaranteed to fail the stringent quality control awaiting them now. For them, it's game over.

If the uterus has not heard otherwise, it will switch to its habitual monthly routine: the mucosa will dissolve and the woman will have her period. New cycle, new mucosa, repeat. It's a troublesome phenomenon, which most mammals are fortunate to escape. The short list of menstruating animals includes humans, monkeys and (don't ask me why) some varieties of bat. But why specifically us? Well, we should probably blame our greedy placentas. Most mammals produce a far more secure variant. In horses, cows and pigs, the cell vesicle sits more or less on the surface of the placenta's mucous membrane; it then winds threads around the mother's blood vessels without destroying them. This gives the mother a good deal of control over what is transferred to her offspring, and a lower risk of serious bleeding if the placenta should become detached. For humans, on the other hand, it was an absolute necessity to create an emergency brake. Allowing you to move in was potentially fatal for your mother. So you had to ask nicely for permission before you could begin sponging off her.

At this point you may have got the impression that we're all little better than gruesome, greedy parasites, invading the bodies of our innocent mothers. It's not exactly a pleasant image, so to correct this, allow me tell you about a fascinating experiment that shows another side of the story. In the darkness of the ocean the jellyfish *Aequorea victoria* resembles a glowing chandelier, thanks to a luminous green protein that it produces. By injecting a fertilised mouse egg with one of *Aequorea's* genes, researchers were able to create male mice with luminous green cells. The scientists allowed these luminous mice to mate with normal female ones, and soon after, the female mice were pregnant. The next thing they did might sound a bit brutal: twelve days later, the researchers gave each pregnant mouse a heart attack. After which they examined the mother mouse's heart, and noticed something quite incredible: some luminous green cells that could only have originated from the baby mice growing in her womb. It appeared that stem cells from the baby mice had found their way out of the placenta and into the mother's bloodstream, and upon reaching the heart they had turned into pulsating heart cells to help repair the damage from the heart attack.

The same thing can probably happen with humans too. Interestingly, pregnant women who suffer heart failure are more likely to survive than those who are not pregnant. When a Spanish research group examined the hearts of two women who had suffered from severe heart failure, they found cells that originated from their sons – despite the fact that it had been more than a decade since they were born. Blood tests have also shown that mothers carry cells bearing their child's DNA for many decades after pregnancy. Researchers have even discovered foreign cells hidden away in the brain. Could there be a tiny bit of you in your mother's body? A single cell beating in her heart, or chatting away to the other nerve cells in her brain? It's nice to think that you were at least a *tiny* bit useful while you were in there freeloading.

NATURAL CLONES AND UNKNOWN TWINS

THESE CELLS THAT burrow and murder their way into the uterine mucosa will never become an actual part of your body. The ones that will become *you* sit hidden within the cell vesicle. One week after conception, you consist of a bunch of stem cells that can form any body part – they could become heart muscle cells, nerve cells, liver cells or anything else. At this stage they are still so flexible that they can even create more than one body. If the cells were to detach from one another, and form two separate cell clusters instead of one, they might develop into two complete people. This is the most common way that identical twins occur, and since the placenta is already being formed, the twins will have to share it. Alternatively, the cells could have fallen apart a few days earlier, when they resembled a microscopic raspberry, in which case two vesicles will attach to the uterus and two embryos, each with their own placenta, will be created. About one third of identical twins begin like this.

Since identical, or monozygotic, twins originate from the same cell, they possess exactly the same DNA strands – they are natural clones. If one of the twins commits a crime, investigators will be unable to distinguish between them using a DNA analysis. However, if their fingerprints were examined, then the culprit would be revealed. This is because fingerprint patterns are in part shaped by the environment in the womb. The two twins occupy different spaces, and therefore experience different streams and pressures against their fingertips. In addition, because the supply of nutrients from the placenta is not evenly distributed, one can grow slightly faster than the other. This means there will be small variations in the twins, even though their genes are exactly the same.

Twins can also result if the mother releases two eggs instead of one, and each becomes fertilised by its own sperm cell. These are called non-identical, or dizygotic, twins, and their DNA strands are no more similar than in normal siblings. But they are not exactly normal siblings. It appears that twins can exchange cells while in the womb, just as cells can be transferred to mothers. In this way, for example, they may end up having two blood groups – one that comes from themselves and another that comes from the twin sibling.

I have no twin that I know of, but perhaps I had one that I was never able to meet. On rare occasions, the two cell clusters recombine before two bodies can be formed. If this happens with non-identical twins, then the child will grow up with two sets of DNA, a so-called *chimera*. Instead of all the cells having the same DNA strands, some of them will carry the DNA strands of the 'twin'. Usually it never comes to light, but the phenomenon can sometimes lead to quite absurd situations. In 2002 Lydia Fairchild, from the US state of Washington, was expecting her third child and had applied for child support. The authorities required that she and her ex-boyfriend take a DNA test to prove they were the parents. As expected, the results showed that the ex-boyfriend was the father. The only problem was that, according to the DNA test, she was not the mother. Fairchild was suspected of fraud and feared that her children would be taken from her. The court summoned a witness who was present at the third child's birth. More blood tests were done. The DNA analysis still showed the impossible: without a shadow of doubt, she could not be the mother of the child she had just given birth to.

How was this possible? Were the tests flawed? Only after taking samples from different parts of Fairchild's body was the mystery solved. The blood and skin samples taken previously matched one another, but the cells they obtained from the cervix were different – they carried a second DNA profile. Fairchild was a chimera. Before she was born, her cells had merged with a twin in the womb. When this happens, instead of each twin making its own complete body, the recombined cells become woven together and share the tasks between them. In this case, the cells that made the skin came from one twin; those that made the egg cells and cervix came from the other. Fairchild's body was created by twin sisters – which made her the child's mother and aunt at the same time.

Unless you have an identical twin, there's not one person on the planet who has exactly the same DNA as you. When the sperm and egg cell merged during your conception, a unique code emerged. But the areas where your uniqueness is manifested are very small – most of the recipe is the same in all people, and these days it's possible to look it up online. Through the Human Genome Project, researchers have mapped the entire human DNA – all three billion letters of it. The mapping was not of any one individual's DNA; many anonymous donors contributed different sections of the code. It was a vast project that took many years to complete, and cost hundreds of millions of dollars. It shows how fast technology has evolved that, today, it costs around $1,500 to do the same thing as an individual (even less if you're content with a rough survey). A laboratory can take some of your spit and provide you with your exact rows of A, T, C and G just a few days later. Altogether, the formula would fill more than a hundred thick books. If you were to read one letter per second, it would take you ninety-five years to finish.

It's unlikely you'd have learned anything more about yourself, either. Imagine a book written without a single full stop, comma or space, and that in some places it's written backwards without warning. The whole thing would be nothing more than page after page of incomprehensible gibberish. Your DNA is something like that – and it's among this chaos, this ocean of apparently random letters, that researchers are now searching for words and sentences that make sense. One of the first things they found was that they had made a serious miscalculation when estimating that humans contained about 100,000 genes. We're not even close to that number. Human beings – the inventors of the computer, founders of civilisations and cities – have only about 20,500 genes each. That's roughly the same amount as the tiny roundworm *C. elegans*. Even the maize plant beats the socks off us with 33,000. In fact, your genes account for less than 2 per cent of your DNA. So what do they actually do?

THE CONTOURS OF A BODY

THE THIRD WEEK

DAY SIXTEEN

(1 MM
ABOUT THE SIZE
OF A POPPY SEED)

AT THE START of the third week, the cell cluster which is soon to become you flattens and spreads out across the vesicle. Right now there's nothing that even vaguely resembles a body – you look more like a little round plate. On each side of the plate are two fluid-filled sacs. One of them becomes the foetal sac, which will enclose you and the little pool in which you'll live for the next few months. The other will become the yolk sac, a round balloon with its cord fastened inside your stomach. The yolk sac creates your first blood cells; a job that your liver, spleen and your bone marrow will eventually take over. When it's no longer needed, the yolk sac will shrivel up and become part of your intestines.

With birds and other egg-laying animals, the most important role for the yolk sac is to provide nutrition –

they have no placenta to feed on, of course – so it's packed with vitamins, minerals, fats and proteins. If you crack open a hen's egg, in addition to the yellow yolk sac you may also notice some thin white threads keeping the yolk attached to the centre of the egg. In a fertilised egg, a chick slowly emerges from a thin white plate on the surface of the sac. At first, it's a barely visible speck, but after a few days red blood vessels coil themselves around the yolk. Soon after that, the yolk sac shrivels up and a living creature gradually emerges. Three weeks later the egg hatches and a new chick is ready to meet the world.

Things go a little bit slower for us humans. But at the start of the third week you take at least one important step forward from the plate stage. Over the course of a few crucial hours, you are given a front, a back, a top and bottom, and right and left sides. It's one of the most critical periods in your whole development. Had something gone wrong, then you wouldn't be reading this book now, with your intestines safely packed behind your skin and your heart beating reliably on the left side of your chest.

The first sign of this dramatic change is that the round plate becomes more of an oval shape. At the same time a thin strip appears. This is the beginning of your back, and it extends from the edge and towards the centre of the oval plate, where your head will pop up later on. If we were to zoom in on this strip, we'd see all the cells wandering down it towards a small pit at its centre. The cells dive into it, forming a new layer under the topmost one. Soon you will consist of two cell plates stacked one on top of the other. Shortly after that, new cells arrive and spread themselves between the two plates, so that you end up with three layers of cells.

This may not sound terribly impressive; I promised you dramatic changes, and all that's happened is that a round plate has become a triple-decker cell sandwich. But you're already infinitely more interesting than the raspberry you were a short while ago. These cells are no longer confused, needy newcomers with no idea where they are or what they're supposed to do. They have completed a rough division of labour. The cells on the top layer will form, among other things, skin, hair, nails, eye lenses, nerves and your brain. From the bottom layer you'll get intestines, liver, trachea and lungs. And the middle layer will become your bones, muscles, heart and blood vessels.

As time moves on, each cell will become more and more specialised. Eventually, you will end up with over 200 different types. Their shape, size and characteristics will vary enormously. Round red blood cells will float around your body, carrying oxygen. Immune cells will patrol for intruders. Your ear will contain hairy sensory cells that dance to every sound you hear, and electrical signals will flicker and spark in your brain through the long threads of nerve cells.

Every single one of these cells will contain exactly the same DNA sequence. This recipe has been copied, over and over, by each new generation of cells, going right back to the moment of your conception. Which poses the question: what makes them all so different from one another?

The answer lies in the proteins they make. Genes don't actually do anything by themselves: they are simply recipes that cells use to make proteins. The cells put away the recipes they don't need and take out the ones they do, meaning that each cell is able to make its own particular set of proteins. The DNA molecules are heavily guarded within the nucleus, like a highly exclusive cookbook. When the cell is about to generate a protein, it first makes a copy of the gene using RNA, a molecule similar to DNA. The RNA molecule then moves out of the cell nucleus and into the cell's protein factory.

Before the cell begins generating the protein, it does some cutting and pasting in the RNA copy; indeed, the same gene recipe is often used in several different proteins. It's like Grandma's apple pie, with a new twist – sometimes she sprinkles almonds on top, sometimes there's a few extra raisins. As soon as everything is ready, the cell's protein factory begins to fuse amino acids, which are the protein's building blocks. The factory reads the recipe, three bases at a time, and the three letters indicate which of the twenty different amino acids it should choose. For example, if it reads GAA then it knows it should attach itself to a glutamic acid. Other letter combinations might be code for a different amino acid, or they can tell the machinery that it should stop.

Eventually, this long chain of amino acids settles into the three-dimensional form of the protein. Depending on the order of the amino acids, the shape they finally adopt could be anything from long fibres to little round beads. There are even some proteins that resemble tiny propellers. Some of the proteins become woven together and form large structures, such as skin or eyelids. Others work tirelessly within the cells, breaking down nutrition, storing energy and the like.

By producing new proteins, the cells can transform themselves and perform new tasks. In the third week, some of them join forces to build your first organs. The cells in the middle layer form a thick string known as the *notochord*. You would keep this string for your entire life, if you were destined to become a lancelet. These fish-like animals lack a skeleton, but the solid notochord prevents their bodies from becoming a limp, gelatinous sausage. Normal fish and humans manage fine without it, though, once the rigid spinal column is finished. In the end, the only remaining part of your *notochord* is the shock-absorbing cushions that sit between your vertebrae. However, while we are embryos, our notochord is just as important to us as it is for lancelets; it sends important signals about what's coming next.

On the signal from the notochord, cells from the top of our three layers begin forming a thick sheet. On either side of the spine, the edges of the sheet begin to fold towards each other, eventually forming a tube, about a month after conception. Later, most of this tube will be transformed into the spinal cord. At the head end, the tube swells up and forms three small sacs – and it's here that the cells embark upon their most ambitious project: the brain. Despite it being one of the first things they start constructing, it is the last to be finished – your cells will continue working on it for quite some time. Even when you are born, it will be far from complete. Researchers used to believe that the brain was more or less finished before puberty, but in recent decades they have found that it is still undergoing major changes right up until the end of our twenties. We'll return to this mysterious organ in a later chapter – but right now we have something much more urgent to attend to. Your innermost cells are getting malnourished. Previously, they got all the oxygen and nutrition they needed directly from their surroundings, but this only works over short distances. As your body grows, your innermost cells run the risk of dying. This might have been the end of you – if not for your heart.

About eighteen days after conception, the cells form two small tubes, one on either side of the spine. Over the next few days, they move towards each other and merge. At the same time, the cells around the new tube change, taking on a very special role: they become heart-muscle cells. These cells spontaneously begin to contract, in and out, in and out. Always. No matter what. In the laboratory, scientists have grown heart cells in Petri dishes and observed that each cell contracts independently. When they come into contact with each other, they begin to beat together – *thump, thump*. So, only twenty-two days after conception, this little heart tube of yours beats for the first time.

On it will go – every day, every second – without a single break. All over your transparent body, tiny red specks appear; merging together to form your first blood vessels. Over the next few hours your cells must constantly build new blood vessels in order to reach out to all the nooks and crannies of your increasingly complex body. These vessels branch out into smaller and smaller byways, the smallest of which are called capillaries. These are so narrow that there is room for only a single, tiny blood cell to squeeze through them. If you were to place ten capillaries side by side, they'd be about the width of a human hair.

Capillary walls are so thin that oxygen and nutrients seep out of them and into all the hungry cells running alongside. When complete, the combined length of all your blood vessels laid end-to-end would reach more than twice around the earth. It's through this colossal network that your blood gets pumped to every cell. But the heart never gets tired. It keeps on going, right up until everything is over.

Nearly all animals have roughly the same number of heartbeats during their lives. Small animals live shorter and have fast-beating hearts, while large animals live longer and have more slow-beating hearts. Humans are a clear exception to this, and live far longer than our seventy beats per minute would predict. Mice, however, follow this pattern typically. A mouse's little heart beats fast and feverishly – at least 450 times per minute – for only a year or two, before giving up. At the opposite end of the spectrum we have the blue whale, the largest animal that has ever existed. A blue whale has blood vessels so wide that we could swim through them, and can live to be one hundred years old. Its heart, weighing more than 100 kg, beats under ten times a minute – each pulse sending a thousand litres of blood coursing through its gigantic body – and is so loud that it can be heard many kilometres away.

Enough about blue whales, though; this story is about you. As soon as your little heart tube begins to pulsate, it pumps fluid around your tiny body. Your cells aren't quite finished making blood vessels or blood, but these little trickles are enough for now. The story can continue, and you can grow bigger than a rice grain. Still, we have arrived at a puzzle: how did the heart actually know that it should appear right there and then? Why did it become a heart, and not a lung or an ear? In order to understand this, we have to look first at how cells talk to each other.

CELL LANGUAGE FOR BEGINNERS

CELLS TALK TO each other constantly. They chat about what we eat and drink, about where bacteria have crept in, about whether we are stressed or afraid. *Should we start an inflammation here? Should these blood vessels expand? Is the heart beating fast enough? Are we breaking down enough fat?* Billions of conversations are taking place without making a sound.

The language of cells is written in molecules. They communicate by sending and receiving chemical messages, often different proteins. Some of them resemble loud cries, and race through blood from one end to another. If you've just eaten, your pancreas screams out a protein: INSULIN! And once the liver cells receive this protein message, they begin to assemble blood sugar into long chains which they save for later. It would be terribly confusing and tiring for the liver if the pancreas didn't keep it updated about your daily meals. Your liver takes care of your blood sugar and has to alternate between storing your energy for later and using it now. If you skip breakfast or suddenly eat a piece of cake before dinner, you can be sure it will be discussed by your cells immediately. The cells can also have more intimate conversations with their neighbours by releasing tiny quantities of a certain substance into the surrounding fluid. In addition, it's not unusual for them to talk to themselves a little. An immune cell that has detected an infection will give itself a kind of pep talk before it is ready to attack.

All cells are surrounded by a thin film called the cell membrane, and only a few molecules manage to sneak through it into the cell without permission. Instead, most of the messages are delivered indirectly by hooking on to a type of molecule called a receptor, which sits on the cell's surface. The messages fit the receptors like a key fits a keyhole. For example, on the surface of a liver cell there is a receptor for insulin. When the insulin molecule hooks on to the receptor, it triggers a chain reaction within the cell, and the liver cells begin to store nutrition.

Many diseases occur due to communication failure between cells. In type 1 diabetes, the pancreas struggles to be heard – it cannot produce enough insulin. For unknown reasons, the body's immune system begins to attack the cells that make insulin, so that the usual Oi! to the liver comes out more like a polite *Ahem*. That means it's up to the patient to supply insulin messages to the body using a syringe. In type 2 diabetes, the pancreas attempts to report that the person has eaten, but the cells don't hear it. The insulin passes through the blood, but the receptors on the cell's surface struggle to pick it up. The danger with diabetes is that the cells are convinced they are starving, no matter how heartily the person eats. The liver, oblivious, keeps breaking down its energy reserves and the blood sugar level becomes dangerously high. Unable to use the sugar, the body has to get rid of it through the urine. In the early days of medicine, it was actually common practice to diagnose diabetes patients by tasting their urine. This didn't seem to bother the English doctor Thomas Willis, who in 1674 wrote that the urine he had sampled was 'wonderfully sweet as if it was imbued with honey or sugar'. He suggested adding the Latin suffix *mellitus* to the disease's name, which means 'honeydew'. The term diabetes mellitus is still used today.

Insulin is just one of many substances that the cells in your body use to communicate. By staying in touch with one another, they become a resilient, well-functioning community, one that contains more inhabitants than there are known galaxies in the universe. You can eat random meals, move between hot and cold rooms, rest, run, get up early or stay awake half the night. Whatever you throw at it, your body will keep everything impressively stable on the inside. It ensures that the blood is sufficiently acidic, it distributes food and energy, clears away waste and chokes bacteria for you, without you ever giving it a thought.

When the cells are assembling your body, they use these chemical messages to share tasks and give each other instructions. No one is the boss, nobody knows what they are doing and no one knows quite what the result will be in the end. After all, you're unique: no one in the world has ever seen what these cells are building. The only thing the cells do is proceed from one stage to the next. The complicated shapes and structures in your body have emerged gradually because the cells have followed a long series of simple instructions. It's a little bit like folding an origami figure. All you do is keep folding step-by-step in different places. You can't see what you've made until you are holding a paper swan in your hands. In nature, we often see that incredible patterns can occur when a group relies on a set of simple rules. One example of this is the patterns made by flocks of birds known as *murmurations*. Every bird is careful to avoid getting too close to its companions while moving in exactly the same direction as them. The result is a spectacular interplay, creating the impression that the birds are following some prearranged choreography.

Build a tube, is the message your cells hear when they start making your heart. When your heart tube first appears, it sits in the middle of a symmetrical body. Your left side looks like a perfect reflection of your right side, and the same applies inside. But it won't stay like that. In the next few weeks, your heart tube forms itself into a compressed S. This runs in a loop and creates four chambers. If everything goes as planned, the finished heart will sit between the lungs, tapering towards the bottom and pointing to the left.

Other organs will also take up permanent residence on different sides of the body: the stomach and spleen will sit on the left, the liver on the right. But how do your cells know left from right? Well, back when you were still a plate, some of the cells along your back grew thin hairs called *cilia*. These hairs began to rotate quickly together in the same direction, creating a stream of fluid that swept towards the left. In this way, they literally guided the conversation in a certain direction. Protein messages sent from cells in the middle of the body were swept to the left. So the two sides of the body received slightly different commands and developed in different ways.

People with the rare genetic disorder Kartagener's syndrome have all their organs on the 'wrong' side. The heart beats on the right, the liver works away on the left. It doesn't sound like this should be a problem, but, in fact, it can lead to respiratory infections and fertility issues, because the small hairs on their cells don't work as they should. The cells use these hairs for more than just swirling up molecules while you're still an embryo. In a finished body, these hairy cells can be found in many places; the lungs, for example, where they sweep away the dust and dirt that you cough up. Without this cleaning function, the bacteria can settle right in and create an infection. The same problem can occur in smokers, because smoking destroys the hairs. A man suffering from Kartagener's syndrome will also have reduced fertility, because the sperm cell's swimming tail functions poorly.

The cells are unable to determine if the message they are getting is rational or crazy. If the heart message is not swept in the correct direction, it doesn't matter if the cells that receive it are actually on the right side. Cells are deaf and blind, they sense their world through molecules. If the message says *heart, heart, make a heart,* then they can do nothing else but obey. But how is it that a molecule can affect the cell's fate?

Believe it or not, we can get the answer from a fruit fly.

THE ART OF BUILDING A FRUIT FLY

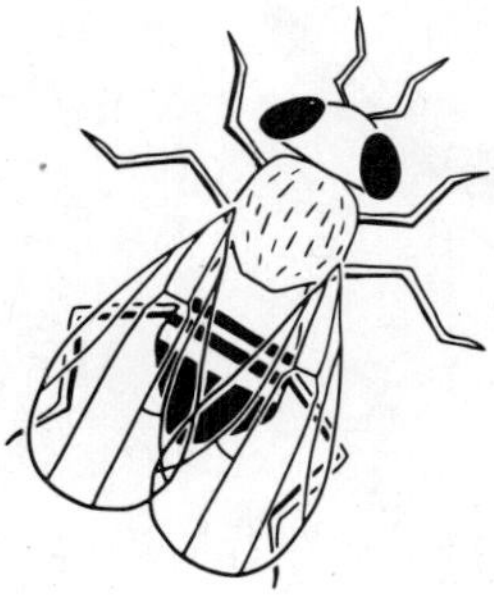

THE FOURTH WEEK

DAY TWENTY-FOUR

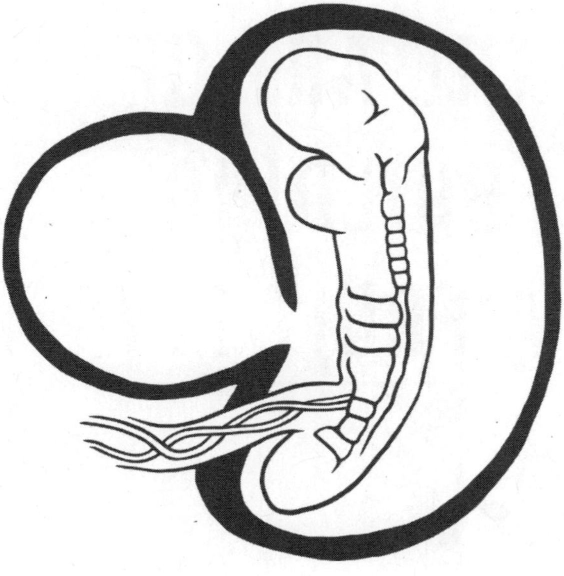

3 MM
(ABOUT THE SIZE OF A SESAME SEED)

WEEK FOUR: TIME we stepped back and admired you a little. You're no longer a single disc. Your cells have wandered, grown, twisted and turned into something resembling a funny little larva. You're only a few millimetres long – and ideally we should slide you under the microscope to see what you really look like – but you have a top and a bottom, a front and a back, and your first organs are growing inside. You also have a pulsating red heart tube, a nerve tube expanding in your head and a bowel tube running through your furry, transparent body.

All this in three weeks. Good going, you might think, except that a fruit fly manages to become a finished larva in less than a day. (To be fair, there's no point hanging around when you live for only a few weeks.) After the egg has hatched, this glistening white larva crawls out to eat and grow bigger. Five days later, its weight has increased over a thousandfold. Now the larva can pack itself contentedly into a pupa, where its cells will apply the finishing touches to this masterpiece. Eyes, antennae, wings and legs appear and, *hey presto,* after about nine days out comes a fully grown fruit fly. Nine days. That's the same amount of time you spent burrowing into the wall of your mother's womb.

For a biologist, the fruit fly is far more than a pest in the kitchen. For more than a century, these insects have been the foot soldiers of genetic research. You really couldn't ask for a better test animal: they're small, undemanding, they live for only a short time and grow rapidly. On the face of it they might not seem much like us. Nevertheless, this tiny larva faces just the same challenge as you and I did: it has to build a body with all the parts in the right places. And to achieve this, you both use the same trick: you divide your bodies into segments.

In the third week after conception your segments are visible for the first time. Two small bumps, called *somites,* appear on each side of your back, near to your head. About an hour later, a new pair appear, then another, again and again until you've made about forty-four pairs, all down your back. Eventually all sorts of strange things will become attached to your spine – you'll get shoulders, ribs and a pelvis – but before your body adds all of this, it creates a repeating pattern. Your spine divides into tiny vertebrae, all with the same basic shape. After that they adjust according to their position: the upper vertebrae will become narrow, allowing you to nod or shake your head; the lower ones will be broader and more sturdy. Your abdominal muscles are also segmented, which is easy to see on a well-trained body.

When the fruit fly is still at its larval stage, the segments are visible as small furrows across its body. Later, the larva will transform into a fly and various body parts will grow out, depending on the position of the segment. The first segments form the head, with eyes and antennae; the middle sections form the thorax, with the legs and the wings; and the last segments become the abdomen. Normally, a happy little fly emerges from the pupa, with correctly placed wings and antennae – but on rare occasions things don't go exactly to plan. Some end up with large, hairy legs protruding from their heads. Others have an extra pair of wings, or feet by their mouths. What went wrong with these creatures?

In the 1970s, researchers began to close in on an answer. The geneticist Edward Lewis and his colleagues at the California Institute of Technology studied the genes of these tiny mutant flies and found that each dramatic transformation was caused by an injury to a single gene. The researchers quickly tracked eight different genes, all of which were found in the fruit fly's third chromosome. Strangely enough, the order of these genes along the DNA strand reflected which part of the body they controlled. At one end of the DNA strand were the genes affecting the head, and at the other end were the genes affecting the abdomen. In between were all the genes affecting the thorax.

These genes are known as Hox genes. If you mess with a Hox gene, the result will be a fruit fly with misplaced body parts. Take the gene Ultrabithorax, for example. Its job, along with the other Hox genes, is to inform the cells that they are in the last of the thorax's three segments. Without Ultrabithorax, the cells would automatically think they were located in a segment further forward, so they would begin creating whatever belongs to that segment. The dutiful cells have no idea that their real task is to actually build the tiny, spoon-shaped balance organs that stick out behind the wings. Which is a shame because, even with an extra set of wings, the fly cannot fly without them. One way or another, the Hox genes ensure that the cells in the different segments behave differently. The question is, how? What are these genes actually doing?

In the 1980s Walter Gehring and his colleagues at the University of Basel attempted to find the answer to these questions. Genetic engineering had made rapid progress, and it was now possible to copy specific pieces of DNA and examine how they were constructed. Letter by letter, the researchers mapped the Hox gene code. Soon they found a string of 180 letters that could fit into any of the genes, whether they organised the body, abdomen or something in between. The researchers knew that the key to understanding how the Hox gene functioned must lie in this string of 180 letters, which they christened the homeobox. But hadn't they seen it somewhere before? They searched their databases to compare the 180 letters with other previously mapped genes and quickly found several hits. What's more, there was a clear pattern: all of the genes made proteins that connected to DNA. In fact, these proteins were known to switch genes on and off, thanks to experiments on an entirely different biology favourite, the one and only *Escherichia coli*.

I get sceptical looks from my friends when I tell them that I cultivate *E. coli* in the laboratory. This unfortunate bacterial family has gained a bad reputation among the general public, thanks to some dodgy relatives who cause violent stomach illnesses. But it's quite unfair, because most of the family is entirely respectable. Harmless E. *coli* already live in your intestines, and actually prevent other dangerous types from moving in. In the laboratory, we allow E. *coli* bacteria to grow in a yellow, highly nutritious liquid heated to thirty-seven degrees Celcius, just the way they like it. In return, the bacteria replicate DNA or create proteins for us. They are our small biological factories and we could not do without them.

In the 1960s the Frenchmen Jacques Monod and François Jacob examined the effects of various nutrients on E. *coli*. They noticed that if the bacteria had access to both glucose and lactose, they'd first tuck in to their big favourite: glucose. (It's like the chocolate box at our house: no one wants to eat the chocolate toffee finger until everything else is gone.) For the bacteria, it's much easier to absorb energy from glucose. The lactose has to be cut into smaller parts by a scissor protein before it can be used, so the bacteria don't bother with it until the glucose has run out. Sensible, but how can such a simple organism carry out such a sophisticated plan?

To create its scissor proteins, the bacteria takes the recipe from a gene in its DNA. First it has to make a copy of the recipe, which it sends on to its protein factory. However, Monod and Jacob found that the bacterium can make another protein which prevents it from making this copy. This protein fastens itself to the DNA string, just in front of the gene, and in doing so it keeps the gene's 'off' button pressed. Only when the obstructing protein detaches itself, can the cell begin using the recipe and make the scissor protein. In addition, they found that the bacterium creates another protein that has the opposite effect: when attached to the DNA strand, it becomes *easier* to make a copy of the recipe. The 'on' button of the gene is pressed, and the bacterium can help itself to the lactose sooner.

This means it's possible to switch genes on and off via proteins connected to the DNA string. And that's precisely how proteins made by the Hox genes function. Each Hox protein connects to its respective part of the DNA string and – *click* – a whole set of different genes is switched either on or off. A fruit fly is far more complex than a tiny bacterium. It is assembled from different organs consisting of specialised cell types that work together. So it must also dedicate large sections of its DNA to controlling where and when it uses its genes. In humans it's even more complicated than that.

Researchers once called all non-genes *junk* DNA because they had no apparent function. Today, this expression is rarely used because we are continually discovering new treasures among these mysterious letter codes. Sitting before and after the genes are codes that act as genetic switches. Certain proteins can recognise these codes and ensure that a gene is switched on at the right time and place. These genetic switches can be compared to light switches in a house. Some of them are multi-connected and turn on all the lights in a room; others just turn on a desk lamp. Hox genes make proteins that switch on whole sets of genes, and ensure that different proteins are produced in the different segments – so that in the end the fly will sprout antennae from its head, and wings from its thorax.

Then comes the big question: what does all this mean to us? I promised that this story would be about you, but now I've written a suspiciously large amount about fruit flies. Where's the connection? Well, we and the fruit fly have a common ancestor (although we need to step back more than half a billion years in order to find it, so we're not exactly close relatives). Until the 1980s it was thought that the genes that organised a fruit fly's body had to be completely different from the genes found in humans. But everything was turned on its head when researchers at Walter Gehring's laboratory began looking for Hox genes in other animals and found them all over the place. They were in worms and fish and frogs and mice as well as humans. (And, of course, fruit flies.) Admittedly, we have four sets of Hox genes, not just one like the fruit fly, but the principle is the same: the fate of the different bumps down your spine is determined by various combinations of Hox genes. They make sure that everything ends up where it should be down your back: shoulder blades at the top, pelvis at the bottom and ribs in between.

Just as it is possible to build tool sheds, mansions and churches with nails and hammers; it is possible to build fruit flies, mice and humans with Hox genes. It is not simply about what genes we have, but how we use them. In fact, we share more than half our genes with fruit flies. Our common ancestor may have been an unassuming worm, but even a worm needs genes to make sure its head is different from its tail. After half a billion years of evolution, those same genes are still in use, but in new ways and in new combinations. As we will soon see, Hox genes are by no means our only souvenirs of the past.

AN HEIRLOOM FROM THE OCEAN

THE SECOND MONTH

WEEK FIVE

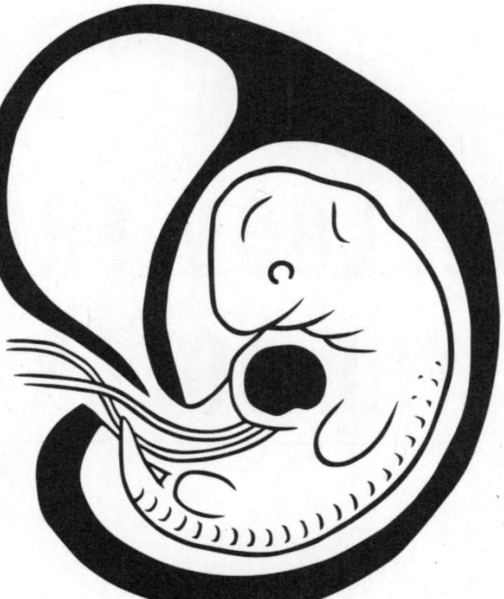

1/2 CM
(ABOUT THE SIZE OF A PEA)

BY THE START of week five you are the size of a pea. Your tiny body is curled and transparent and your head is bowed down towards a long tail. You don't have a proper face yet, just faint outlines of eyes. It's impossible to see that you're going to become a human being; you look more like a prawn. You've also developed four small folds along your neck, separated by deep creases. Just below them is the pounding red lump of your heart.

Previously, everything your cells have done has seemed quite logical: they have grown and folded, made sure everything is in the right place and built a foundation they can build on. But now it looks like they have messed up. Why, for example, are they building this totally unnecessary tail? It will end up being no more than a bony stump that hurts when you fall on your rump. No engineer in his right mind would do anything similar. And what's the point with these folds on the neck that disappear later on? Don't they look suspiciously like what fish reshape into gills?

We humans are not alone in taking such detours. If you study the embryo of a lizard, a chicken or an elephant, you will always find the same strange creature. At the beginning of the nineteenth century the German biologist Karl Ernst von Baer noticed these similarities but he didn't know how to explain them. Then Charles Darwin came along with the answer. In 1859 he published *On the Origin of Species*, in which he devoted an entire chapter to embryos. He explained that the cause of the mysterious similarities between the embryos of widely varying species must be that they all have common roots.

We have a common history with fish, salamanders and chickens, a history that spans hundreds of millions of years. Our early ancestors resembled fish, and roamed the vast primitive ocean. Later, moss began to carpet the barren rocks on land, and scorpions and centipedes emerged to explore the increasingly green undergrowth. Plants grew larger and thrust deep roots into the ground, which was now covered with fertile soil. Soon it was swarming with insects, among ferns as large as trees. The air was warm, humid and packed with oxygen. Our ancestors swam around the swamps of the primitive forest. Later, some of their descendants grew lungs and thick fins. About 400 million years ago, the first amphibians crawled out and on to dry land. But the amphibians stayed close to the water, because that's where their lives begin; without water, their eggs would dry out and shrivel like deflated balloons. Later still, reptiles came along with a solution to this problem: they wrapped their eggs in a protective membrane, which prevented them from drying out. Then, around 200 million years ago, the first mammals arrived, whose offspring were allowed to grow within the safe surroundings of the womb – and gradually a naked, two-legged mammal known as the human evolved.

The last ancestor that we share with chimpanzees lived approximately six million years ago. We humans have only existed for a brief moment in life's 3.8-billion-year history. But we too begin our lives as aquatic creatures: we create our very own salty ocean of foetal water and stay there until we are ready to draw our first breath.

If we just remember that our body is essentially a reconstructed fish, there's a lot we can forgive it for.

Things that seem stupid and illogical suddenly make sense. Take hiccups. You were perhaps only a twelve-week-old foetus when you first encountered this mildly annoying experience. When you hiccup, your breathing muscles suddenly contract, so that you inhale sharply. Then your vocal cords instantly close, and the characteristic *hiccup* sound occurs. The phenomenon is most probably inherited from our amphibious forebears. Halfway through its development, a tadpole possesses both lungs and gills, as our ancestors once did. When it breathes underwater, the action is similar to an elongated hiccup. By closing its throat, it blocks access to the lungs and pushes the water out through its gills. Vitally important for a tadpole, but a quite useless reflex for humans.

Another example is the small vertical groove that runs from your upper lip to your nose. It has a nice medical name – *philtrum* – and when I was small, I thought it was intended for collecting nose drips. But it doesn't really have any special function. It is just a result of the cumbersome way our faces are formed. Your face begins as three separate parts; your eyes sit on either side of your head just like a fish's, and your nostrils sit right on top. Then each part begins to move around slowly, towards the same point; your nostrils creep down from your forehead and your eyes move round towards the middle. Eventually, all these parts converge on a point just below where your nose is today. It is absolutely crucial that the three parts meet at the same time. Even a slight delay at this point might leave clear traces: you could be born with a cleft palate. But if everything goes to plan, then your skin and muscles will be seamlessly woven together. The only evidence remaining will be the little vertical groove – a reminder of a time when you looked completely different.

The four folds and splits we get on our necks while we are still embryos are also an heirloom from the ocean. In a fish embryo, these splits form the gaps between the gills through which water flows as it breathes. The uppermost folds form the fish's jaw, and the last two form the supporting tissue for the gills. Amphibians, reptiles and mammals develop the same folds and splits, but evolution has found new uses for them in each case. If you studied amphibians and reptiles, you would find that the second fold is the source of a small ear bone not found in fish – the stirrup. This tiny bone – shaped, as you might imagine, like a stirrup – makes it possible to hear sound in air. Fish have it easy: when sound waves pass through water, they send vibrations through the fish's entire body, which in turn reach the hearing organs located just behind its eyes. On land, sound waves need to be amplified before they can activate our sensory cells. The vibrations cause the stirrup bone to strike the membrane bordering the inner ear, sending waves through the fluid behind it and causing the hairy sensory cells within to dance to the tones. Some cells prefer the tight, fast waves of the high tones. Others prefer dancing to the broad, low ones. In each of these dancing cells, small channels open, transferring chemicals to a nerve cell which sends electrical signals along thin threads, first to the hearing nerve and then to the brain.

Mammals have taken their hearing a step further. By studying fossils, scientists have seen how bones at the back of a reptile's jaw have become smaller over time, until they finally ended up inside the ear of one of the first mammals. These two bones are called the hammer and the anvil; they help the stirrup to amplify sound waves, by vibrating in sequence, behind the eardrum – until the signal finally reaches the inner ear. It is thanks to this remodelled jaw that we have much better hearing than reptiles.

It is not only in fossils and embryos where researchers find traces of our evolutionary history. Today we have a new tool at our disposal: we can compare DNA.

Darwin, who didn't know how genes and heredity worked, would have surely applauded us for all we have discovered. We have seen that fruit flies, fish and humans have all inherited important genes from a common ancestor. We use these genes to organise our bodies into a basic form, with a front, back, head and tail. From a more recent ancestor we have acquired genes that have allowed us to build a skeleton, a spine and a brain.

Humans, birds and fish look entirely different at first glance, but use the same genes to construct their bodies. How is it that these particular genes have been preserved when so many others have changed? Possibly because they are so extremely important. A gene that starts working early in the process is riskier to tamper with than one that kicks in later. It's the difference between tearing down a load-bearing wall in your house and adding a porch to the front. Embryos with defects in these important genes will perhaps never become fully developed, and therefore have no chance of passing the altered gene on to their children.

So it is easier to refine the details, and add new characteristics, bit by bit. We just have to accept that the road to becoming a human is a little winding.

FLESHING OUT THE PLAN

YOU WILL ALWAYS carry your inheritance from the ocean, but it will soon become clearer that there is a human being growing in the womb after all. At the start of week six you are now about a centimetre long. The folds along your neck are beginning to merge into a face, and your eyes are now visible as two dark specks. Your head leans against your chest, from where your red heart bulges out; and you still have a long tail, but it has stopped growing and will soon disappear. Your brain tube and blood vessels are visible behind a thin, transparent layer of skin; and on each side of the upper body, and down by your tail, you have grown little shoots that will eventually become your arms and legs. The same goes for chicken's wings or the massive legs of a hippopotamus: they too start off in exactly this manner. Even whale embryos exhibit similar looking shoots, despite the fact that they grow neither arms nor legs.

Whales, in fact, originate from mammals that walked the earth on four legs before returning to the sea. Their closest living relative is the hippopotamus. At first, whales follow the same construction plan as other mammals. But by the end, their forelimbs have turned into flippers, and all that remains of the hind-limbs are a few small bony stumps. Your tail will meet the same fate. It will shrink gradually, until finally all that remains is your tailbone. However, unlike the whale you definitely do have the need for a set of arms and legs, and soon your shoots will grow outwards, until they look like tiny paddles.

In these paddles your cells begin work on a first draft of your skeleton. It is made of cartilage, a solid material consisting of cells, protein fibre and a shock-absorbent jelly. First the cells build the forerunner of your upper arm bone. As your arm grows out, the cells move on to the forearm bones and will finally make your fingers. Your legs are built in just the same way – from the innermost to the outermost parts. In order for the cells to build the right bone in the right place, they have to know exactly where they are, and this information comes to them from different doses and combinations of chemical signals.

One example of this is the protein Sonic Hedgehog. The name may sound familiar, and if you're wondering how a protein can end up being named after a video-game character, once again the answer lies with the fruit fly. In order to figure out what a gene does, geneticists often research what changes occur when the gene *stops* working. So it has become a habit to name any newly discovered gene after what happens to the test animal when they destroy the gene. When geneticists studied fruit fly embryos in the early 1980s, they found that when a particular gene was destroyed, the embryos were covered with small, pointed growths. It reminded them of a little hedgehog, and so they called it the Hedgehog gene. When researchers discovered three variants of this gene in humans, they decided to name two of them after different hedgehog species – Indian Hedgehog and Desert Hedgehog – and the last one after the popular gaming character, Sonic.

The Hedgehog genes are certainly not the only ones to be given an odd name by researchers. Take the Ken and Barbie gene, for example. Fruit flies with a mutation in this gene lack external genital organs – just like the dolls the gene is named after. Another example is

Swiss Cheese. If the fruit fly is unfortunate enough to get this broken gene, it will develop a brain that – like Swiss cheese – is full of holes.

When your cells constructed your body, they used Sonic Hedgehog in many places: the intestines, lungs, brain and hands – to give just a few examples. The reason the same message can be reused in so many different places is that the cells interpret it differently. How the cell responds to the message depends on what it has experienced earlier, how large a dose it receives and when it receives it. In principle, it's exactly the same as we humans speak to each other. We can interpret a sentence in completely different ways, depending on the situation. If a colleague were to approach me at the laboratory one morning and ask if we should do some experiments together, I would probably be pleased. If some sleazy guy came up to me in a bar and said the same thing, I would be a little more sceptical. Weirder still, if someone screamed at me, over and over again, 'COME ON! DO SOME EXPERIMENTS WITH ME!' that would be too large a dose in either situation.

So, how do the cells in your little paddles react to Sonic Hedgehog? The protein message is created at the spot where your little finger will later pop up. It then spreads out to the surrounding cells like milk in a coffee cup. Near the source, where there are masses of Sonic Hedgehog, the cells understand that they have to make a little finger. Further away, where there is less Sonic Hedgehog, they form a ring finger, middle finger and index finger. The cells that receive hardly any Sonic Hedgehog, perhaps almost none, make a thumb. In this way the same message determines several different fates at once.

At first your fingers are webbed, but in week eight they become separated. This transformation occurs through well-coordinated mass cell suicide. It all starts when your future finger cells send out death signals. As soon as its neighbours receive these signals, they begin to degrade the proteins. The DNA strands – which the cells would normally do everything to protect – get sliced up into tiny pieces by scissor proteins.

Everything gets destroyed. In the end, there is nothing but crumpled bags of remains left behind. Scavenger cells wander around mopping up the waste, so that the space between the fingers is cleaned up and gradually – one cell death at a time – the paddle becomes a living hand.

Towards the end of week seven, your toes begin to appear. Your long tail is almost gone, and the wrinkles on your face begin to smooth out. You've grown a short, flattened nose, two little ears – and pointy elbows and knees stick out from your short arms and legs. Your skeleton remains a cartilage prototype that will not begin to be replaced with proper bone tissue until the third month, and even then it will be a long ongoing process. When you are born, your bones will still be quite soft, allowing you to squeeze through the birth canal. Your kneecaps will remain cartilage until you are three years old, and your skeleton will continue to develop, replacing cartilage with bone, until you are in your twenties.

When the cartilage begins changing into bone tissue, the cells at the centre of the future bone begin to swell up into giant cells. Shortly afterwards they will die, leaving a cavity that will become bone marrow. The liver and spleen, currently working as temporary blood producers, can soon breathe a sigh of relief. As your birth date approaches, the new bone marrow takes over their job, and will continue with it as long as you live. It is no easy task either. If you start bleeding, stem cells from your bone marrow will change into blood platelets and close the wound. If you get sick, your bone marrow will send a squad of fresh white blood cells which swallow or poison the bacteria. When oxygen is in short supply, red blood cells are soon on the way. In addition, bone marrow works to replace blood cells that are just burned out. Every single second, in every single person, about two million red blood cells say *thank you and goodbye*, and just as many new ones need to be ready to take over immediately.

After the bone marrow cavity is formed, the surrounding cells begin to turn into bone cells. They convert the surrounding jelly into a hard mineral mass. Calcium and phosphate crystals stick to the protein fibres, creating a material that is both strong and elastic, perfect for absorbing shocks without fracturing. Since nutrients are no longer able to float in, thin tendrils stretch out from the bone cells through small ducts and connect to the blood vessels. This allows the cells to continue eating and breathing, despite being hidden among lifeless minerals. As long as you are alive, your bones are alive. Your bone cells make adjustments and replacements every day, which means that every ten years, roughly, your entire skeleton will have been replaced. And while some bone cells build new bones, other types eat up the old ones. After digesting their meal, they release calcium into the bloodstream. Usually the bone-eating cells and bone-building cells work at the same rate, so that you don't actually lose any bone mass. But this system can occasionally go out of sync – as NASA knows very well.

After only a few days in space, an astronaut will begin to lose bone mass, blood calcium level will rise and the risk of kidney stones will increase. This change is probably due to the bones adapting to how they are being used. Because of the weightless conditions, astronauts are barely exposed to any strain, and their bodies consequently slow the production of bone-building cells. The bones optimistically adjust to their new environment. (How would they know that the astronaut doesn't intend to float around for ever?) At the same time, the bone-eating cells continue to feed themselves as usual, and as a result the bones become gradually more porous and susceptible to fractures. Researchers have observed similar effects on the skeletons of people who've been bedridden for long periods. Exercise has the opposite effect: the increased strain on the bones makes them stronger and more solid.

Several of the body's processes depend on calcium, and bones act as a storehouse for this mineral. If the heart or nerves scream for calcium, the bones are quick to sacrifice themselves. After all, it's better for you to have slightly porous bones than a stopped heart. The bone-eating cells do their work, and send calcium into the blood, where it gets passed on to all who need it.

For a while you can manage just fine with a cartilage skeleton. You are floating freely in the womb behind a translucent membrane, like an astronaut in space. During week eight your new arms and legs begin to make small reflex movements. Toes and fingernails begin to take shape, and your ribs stand out clearly in your slender body. Your skeleton and blood vessels are clearly visible under your skin, which is still thin and transparent. After this week you'll no longer be called an embryo, but a *foetus*. You have now made the basic forms of all your organs, but there is still a lot to be done before you can be born.

SEX AND SEA WORMS

THE THIRD MONTH

WEEK NINE

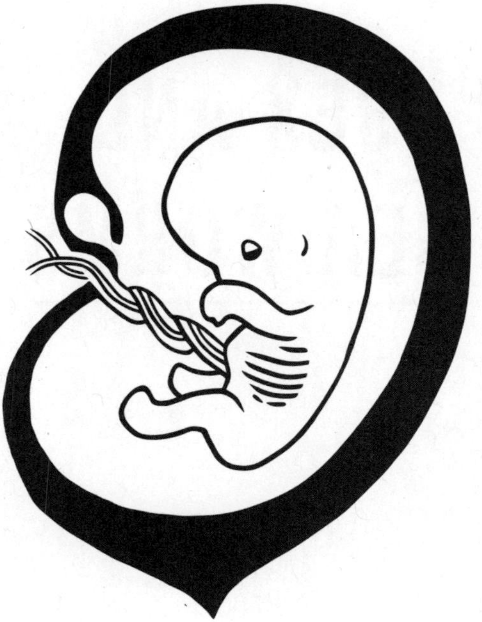

BY THE START of the third month you're about as big as a strawberry. Your nose is wide and blunt, your eyes are far apart. Your high brow and large head make you look a bit like a levitating alien, but over the next few weeks you will acquire more human features. Your dark eyes will be covered by thin eyelids, your bowed head will straighten up, your chin will grow and you will get a more distinct neck.

During this month it will become possible for the first time to see if you are going to be male or female. For the first few weeks there are no differences between the sexes, and this is probably why boys have nipples – not because they need them, but because they are already in place before the sex differences stand out. Even the inner genitalia are made from the same fundamental structure. Regardless of its sex, every embryo will form two sacs, each of which is connected to two small ducts. Then, during week seven a transformation begins, and your genes will decide what happens next: if you have a Y-chromosome in the last chromosome pair, these sacs will become testicles. If you have two X-chromosomes instead, they will become ovaries.

The Y-chromosome itself looks quite forlorn. By comparison with the X-chromosome, which is found in both women and men and which contains between 800 and 900 genes, the Y probably has as few as 50 or 60. In early development, an embryo destined to be female will turn off one of the two X-chromosomes permanently. This is necessary in order to prevent the cells from making a double dose of everything contained in the X-chromosome. The more copies of a recipe the cell has available, the more cooks will become involved, and the more the end results will pile up. When the X-chromosomes are turned off, the embryo already consists of many cells. Precisely which X is chosen is random, which means some of the cells will use the X-chromosome inherited from the mother, while others will use the X-chromosome inherited from the father. Because of this, all females become a genetic patchwork of sorts. In cats, the consequence of this is particularly visible: the gene affecting hair colour is on the feline's X-chromosome; so females can have fur that is a patchwork of colours and patterns. Some of the cells make colour pigments with the recipe they inherit from the father, while others follow the mother's recipe.

The Y-chromosome in men carries a decisive gene called SRY. Without it, the cells would automatically build ovaries. The protein made by the SRY gene doesn't do much on its own, but it acts as a switch that turns on many other genes spread around on different chromosomes. Together, these genes kick off the construction of the testicles, which after a while begin to send hormones into the foetus. The first hormone they send causes one of the ducts, connected to the testicle, to be reconstructed; in women this duct remains unchanged and later becomes the ovary and uterus. The second duct, which runs from the testicle, stays as it is – and will soon be used as a sperm duct, among other things. A little later, the testicle cells begin to produce large amounts of testosterone, which in this context is like saying *Turn into a male!* The message spreads all around the body, and soon after that the differences in sex become visible.

Researchers have carried out experiments on rabbit embryos, during which they removed the reproductive glands at an early stage. Consequently, the embryos developed the bodies of female rabbits even if they had a Y-chromosome. In both rabbits and humans, the cells outside the testicle will never double check if they actually do have a Y-chromosome. It is up to the testicle to tell the rest of the body that the embryo will be a boy. If the other cells do not hear from the testicle, they will build a female body.

With a system like this, it's no wonder that misunderstandings occur. What happens if the cells never hear the testosterone blaring from the testicle? The cells have receptors on their surface which catch these messages and pass them on to the interior. If the receptor doesn't work, then the testicle is producing testosterone for no reason. The cells can't hear it, and so they begin to form a body with female features. On the outside, people with this disorder may appear to have female genitals, since the fate of the external sex organs is determined by the testosterone signal. On the inside, however, this person will have a gland that behaves like a testicle, and the ovaries and uterus will be missing because the duct that forms them has been destroyed on the testicle's orders. To put it another way: sex development is a complex process and it depends on far more than just the Y-chromosome.

Not all animals allow chromosomes to determine their sex. Alligators use temperature to decide everything. If an alligator egg is exposed to temperatures below thirty degrees during the first three weeks, the embryo will become a female. If it gets warmer than thirty-four degrees, a male alligator will grow in the egg. If the thermostat lands in the middle, you'll get a mixture of male and female young, with a preponderance of females.

An even stranger way of determining sex is used by a sea worm called *Bonellia viridis*. This worm begins life as a tiny sexless larva, which for a while floats around in the ocean before eventually sinking to the seabed. Precisely where it lands on the seabed is absolutely crucial. If the larva lands on an uninhabited area, it becomes a female, roughly ten centimetres long. It's hard to describe accurately what a female *B. viridis* looks like, but try to imagine an alien with a body resembling a gherkin and a tail that looks like sea grass. This creature spends the rest of its life attached to the seabed, where its diet comprises the remains of small animals and plants. There's a quite different fate waiting for the larva that lands, not in a vacant part of the seabed, but on the skin of a female *B. viridis*. This larva will be transformed into a tiny male, between one and three millimetres long. Then he'll crawl inside the female's body and spend the rest of his life there as her personal sperm donor. In return, she will ensure that her mate is provided with some of the food she catches. Of all the relationships found in nature, this must surely be one of the most intimate.

There are also some animals that can change sex to suit their environment. Take *Thalassoma bifasciatum*, a fish living in the coral reefs of the Caribbean Sea. Should a male fish move into a reef that is guarded by another male, it would be extremely reluctant to steal his place. Much easier just to turn into a female and live happily beside the other female fish in the small coral community. If the patriarch should at any point die, he is quickly replaced: one of the female fish (usually the largest) almost immediately turns into a male. It takes just a day for its ovaries to shrivel up and be replaced by testicles, thus ensuring the future of the community.

Things are decided for humans much earlier in the life cycle. If a foetus has a Y-chromosome and the signals reach where they should, it will begin to develop a penis. The penis grows from a tiny bud, which in girls becomes a clitoris. Roughly three months after conception, this bud will have grown large enough for the sex organs of the foetus to be visible externally. But the testicles of the male foetus remain for now inside its body, where they will stay until month seven. Then they will first be drawn slowly down to the stomach before making their final journey to the scrotum.

We can probably blame our ancestors in the primitive ocean for this slightly cumbersome process. Fish actually keep their testicles right by their hearts throughout their lives, which might be fine for a fish but is no use to a human. Sperm cells do not thrive in too warm an environment. Fish, which are cold-blooded and change temperature according to their surroundings, manage fine with their testicles tucked away deep inside their bodies. Humans, on the other hand, have testicles outside the body to keep the sperm cells away from the hot internal organs. The little bag they live in can contract or expand depending on whether it is hot or cold outside, and this ensures that the sperm cells are always at their optimum temperature.

SECRET PREPARATIONS

YOUR CELLS MIGHT be having a busy time in the womb, but it's a spa-hotel compared to what's in store after you are born. The heat and the cold outside are things your cells don't have to worry about – your mother keeps things at a comfortable, and stable, temperature of thirty-seven degrees. And thanks to the placenta being full of nice warm blood, your cells never need worry about whether there is enough air or food. Later in life, however, your cells will face a heap of new challenges: baking-hot sun, running, forgotten water bottles and salty crisp crumbs. So, be glad that you acquired a pair of kidneys when you had the chance.

The kidneys and urinary tract are made at the same time as the genitals, and originate from the same cell vesicles. Like so many other parts of the body, they are made in a wonderfully convoluted way, and one of the clearest examples of how haunted we are by our ocean-dwelling past. The cells build things, remove things, change things round and change their minds. It took three attempts before they managed to make the kidneys sitting inside you right now.

What happened to the abandoned drafts? The first were a collection of small tubes that formed right by your neck during the third week. Unfortunately, these primitive kidneys were useless and quickly disappeared. A little later, a new pair appeared further down your back. These sausage-shaped kidneys looked quite similar to the ones found in fish and amphibians, and you actually used them in the womb for a short while. (In girls this pair also disappear, but in boys some of the cells stay behind and become part of the genitals.) Then, in week five, construction of the final kidneys began. However, just to make it extra difficult, they popped up in the wrong place, meaning that they had to go on a journey before they could settle down. First they moved down towards your pelvis and attached themselves to your bladder, later they turned round and moved upwards. Eventually, they arrived at their final position: on each side of your spinal column, at roughly the same height as your lowest ribs.

The finished kidneys are a reddish-brown colour, bean-shaped, each one the size of a clenched fist. Their normal working day looks something like this: receive blood, clean it, pass it on, repeat 399 times – *Phew!* Each kidney is composed of many small channels that are connected to coils of blood vessels. The fluid is extracted from the blood, and then flows through the channels, where the kidney filters away what you don't need and returns the rest. Just as the potato peelings and discarded packaging accumulate when you make dinner, a certain amount of rubbish is also created and needs disposing of when your cells are at work. For example, you need to remove the ammonia formed when you break down proteins. If ammonia builds up in your body it is extremely toxic. Fish have the same problem, but for them ammonia is easy to get rid of – they just release it straight into the water. For land-dwelling animals, it's not a good idea to walk around peeing constantly; access to water can be limited, and we have to preserve the fluids we have as best we can. So the liver helps us to quickly convert the ammonia to urea, a substance we can tolerate in higher concentrations. After that, the kidneys can sort out the urea and send it to the bladder through a tube; and then, when the time comes, you can get rid of it by urinating.

Despite everything we do, our kidneys help keep us remarkably stable on the inside. They carefully monitor the amount of water and salt in our bodies, because if the salt concentration gets out of balance things can go very wrong.

Your heart cannot beat without salt, because you need it when your muscles contract. Nor can you think or feel without salt, because the nerve cells use it to create electrical signals. In short, without salt you would be stone dead. There are salts in your blood, in the fluid that surrounds the cells and inside each cell itself. Salt-water spray that you can buy to hydrate or clean your nose contains 0.9 per cent salt to emulate the fluid surrounding the cells inside your body.

If you place a cell in a solution with too little salt, it risks bursting like an overinflated water balloon. Nature especially likes to even out differences – a cell that is saltier than the water outside is something it cannot accept, so the cell begins to absorb water to dilute the salt. Conversely, if the cell is placed in a solution that is saltier than itself, the opposite happens. The poor cell has to donate water to its surroundings, whether it wants to or not, and eventually it shrivels up into a limp, wrinkled blob resembling a raisin. Bacterial cells are no better at dealing with too much salt, and this is one of the reasons salted food lasts a long time. You can thank your kidneys that you're not just a few thousand billion floppy raisin cells.

Anyhow, as long as you are connected to the placenta, you can relax. You can simply pass all of your waste to your mother's blood and let her kidneys do the job for you. At the same time, your own kidneys can start rehearsing for their future job in peace and quiet. In week nine, your kidneys will already start producing urine. One week after that, you will start drinking small amounts of foetal water and peeing it out, and you'll continue doing this until you are born.

That you spend several months bathing in your own urine and – as if that wasn't enough – even drinking from this polluted swimming pool, sounds rather disgusting. What on earth can this be good for?

Quite a lot, as it happens. What you are actually doing is both vital and ingenious, and nowhere near as disgusting as it first sounds. Your mother makes sure your watery home is regularly cleaned. The walls that separate you from the placenta are like a leaky sieve, full of tiny holes through which molecules can slip. The waste sifts out of the foetal water and into your mother's blood, so that every three hours all the liquid content has been filtered. Not only is this system good training for the kidneys, but also excellent preparation for one of the most important things you will do after you are born: drinking milk. Over the next few weeks your sucking muscles develop and you get chubbier cheeks. At the same time, your intestines are busy absorbing nutrients from the foetal water. You are slowly preparing yourself for life on the outside.

THE BRAIN'S INNER WANDERINGS

THE FOURTH MONTH

WEEK THIRTEEN

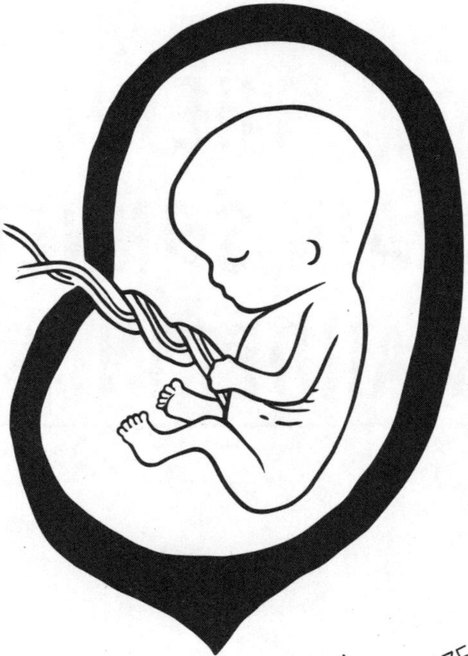

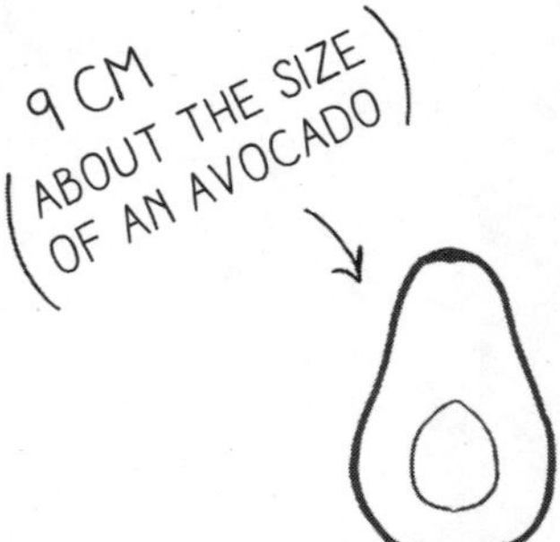

AT THE BEGINNING of the fourth month, you are about as big as an avocado. Your head is upright, your throat and neck are distinct, and a network of red blood vessels runs under your thin skin. You are about to become an active little creature, one that rolls over and jerks and kicks. Sometimes you might stretch your arms out slightly, or suck your thumb. (By the way, the thumb you choose to suck is no casual matter: most of us prefer the thumb on the right hand, but left-handed people often choose the left thumb before they are born.)

Over the course of this month, more and more of your jelly-like cartilage skeleton changes into bone. Meanwhile your body grows rapidly, your legs catching up with your arms, which already had a head start. Your head is still disproportionately large, and inside your skull work continues on the most intricate organ of all.

A lot has happened since the time your brain consisted of just three small sacs at the end of a tube. These vesicles have grown, curled up and divided. The one furthest back forms the cerebellum, among other things, which is important for controlling movement. The same vesicle, along with the central sac in the trio, also forms the brain stem, which regulates breathing, heart rate, sleep and other basic bodily functions. The foremost pouch divides into two halves that grow quickly, spreading out over the other parts of the brain and hiding most of them. On its outermost surface comes the cerebral cortex, which is responsible for the most advanced brain functions. Humans have an exceptionally large cerebral cortex, which is why we can make calculations, philosophise and read and write books. Eventually it grows so large that it must crinkle up in order to fit, but this doesn't happen until the very end of foetal development. During the fourth month it is still smooth, like the brain of a mouse.

At this point around 200,000 new nerve cells are popping up every single minute. Deep within the brain your stem cells are dividing over and over again. After each division, one of the cells stays put while the other sets off on a winding journey towards its new home.

Just like a backpacker after finishing college, the nerve cell uses this journey as a means of finding itself. *What kind of nerve cell should I really be? Should I work with sight? Motion? Smell?* It picks up signals from other cells it encounters on the road, which helps turn the correct genes on and off.

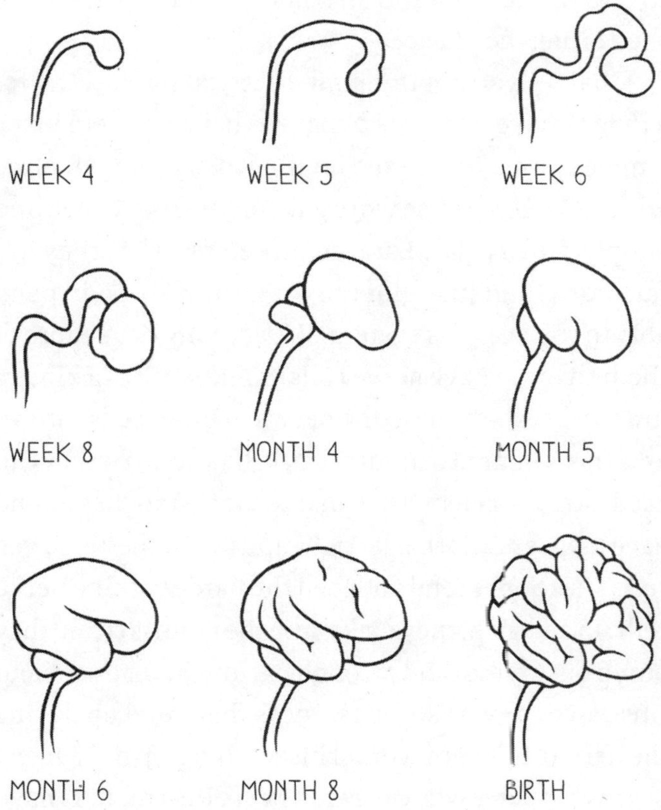

The later a nerve cell is formed, the longer this journey will be. That's because the brain constructs itself from the inside out, layer by layer, which means the deepest, most primitive areas of the brain are those formed first. As a result, and as the brain grows bigger, the nerve cell's journey becomes increasingly difficult to undertake alone. So another type of cell comes to the rescue: the glial cell.

Glial cells do not transmit electrical signals, and for a long time researchers believed that they were just a simple connective tissue holding everything in place within the brain. The word *glia* comes from Greek and simply means *glue*. Later it turned out that they are far more than just glue: in fact, they are indispensable to the nervous system. We actually have more of them than we have nerve cells. Some work as a kind of immune system, crawling between brain cells into an area that's injured or under attack. If necessary, they can eat destroyed cells. Other glial cells are star-shaped and have long tendrils that lie right up against the blood vessels. These ones tend and feed the hard-working nerve cells. In addition, they make sure the brain is clean: they pump out excess fluid through narrow water ducts, and rinse away any waste substances that build up during the nerve cell's hard work. This washing-up most likely begins properly while we are asleep. Researchers have observed that the brain's pumping system is ten times more active in sleeping mice, and that some cells shrink at night to make way for the stream of fluid. This means that, every morning, we can start the day with a freshly washed brain, thanks to our glial cells.

Initially, there's a special type of glial cell that helps the newborn nerve cells on their way through the expanding brain. The glial cells sprout their long tendrils right through the brain layers. Generation after generation of nerve cells latch on to them and haul themselves upwards towards their goal, like tiny snails each moving up a blade of grass. In the end, the nerve cell finally reaches its new home – but that is when the real challenge begins.

As soon as it arrives in position, it must do as most new arrivals do: get a network. If there is one thing nerve cells live for, it's talking. Your brain is teeming with nerve cells deep in excited conversation, some with over a thousand others at once. Other cells sit in your skin or deep inside your nose, transmitting messages about everything you sense. Your spinal cord is packed with nerve cells. The cells in your spinal cord work closely with the brain and talk frequently to your muscles. Wiggle your toes and thank your spinal cord that the muscles down there understood what to do.

When nerve cells talk, they do it in their own very special way: through electrical signals. While hormone messages drift slowly through the blood and can be picked up by unauthorised recipients, the nerve signal is lightning fast – and private. Long fibres sprout from nerve cells in all directions: these are the literal wires of your nervous system. The nerve cell's main communication line is called an axon and it transports information away from the nerve cell. In order to keep your nervous system from turning into a disastrous cable-salad, it's important that the axon connects to the right place. The nerve cells working with vision have to connect to the eye, and those controlling motion must connect to a muscle. If you want to wiggle your toes, the nerve cells at the bottom of your spinal cord must produce axons that extend all the way to the muscles in your feet. Which means that these nerve threads need to be over one metre long in an adult human. How on earth do they get to where they want to go?

Luckily, unlike the wires of a computer, the wires of nerve cells are curious, inquiring and entirely alive. The axon creeps through your body, guided by molecules on the surface of the surrounding cells. The thin nerve fibre stretches forward, testing them: *Can I hook on to anything here?* Then it stretches a little further and finds new molecules to connect with. Axons from different nerve cells prefer different surfaces, and therefore choose their very own microscopic paths through the chaotic cellular tissue. In addition, the nerve cells are guided by attractor molecules that spread out from the target, just like sperm cells swarming around an egg. Right at the tip, the axon forms a fan shape comprising numerous thin tendrils; this serves as a kind of sensor that it uses to sniff its way towards the right spot. Some axons grow out from your eyes and into your brain. Others grow down your undersized legs and send the commands that make you deliver your first kick.

When the main wire finally reaches its target, it picks up a protein message and sends it back through the nerve fibres to the core of the cell. That protein molecule switches on the genes that will secure the cell's future. If the message doesn't arrive in time, the cell will assume that it never succeeded in connecting to the target and commit suicide, shrivelling up into a floppy bag just like the cells between your fingers did. In fact, this unhappy fate awaits many of your nerve cells because you produce too many of them. The nerve cells are in competition with each other, and only those that make the best connections survive. Many nerve cells have sacrificed themselves to make your brain as good as it is.

The nerve cells lucky enough to succeed form a tight bond with the glial cells. A special type of glial cell wraps itself around the nerve cell's axon and packs it in a fatty substance called *myelin*. Just like the plastic coating on a wire, this fat layer acts as insulation and ensures that the electrical signal doesn't leak away.

Glial cells begin by wrapping round the axons in the spine and the innermost, oldest structures of the brain, working their way slowly outwards. The whole process lasts for years, and the outermost areas of the brain are not fully insulated until you are in your late twenties. The final area to be completed is the prefrontal cerebral cortex, which is important for personality and your ability to plan and assess the consequences of your actions. With that in mind, it's perhaps not so strange that a teenager might have trouble understanding that tequila shot number five was a bad idea.

Our brain's most dramatic changes take place by the time we reach thirty, but it will never be completely finished. Your brain is your life project. Everything you learn and remember leads to physical changes in the connections between your nerve cells. By the time you are finished reading this book, your brain will be slightly different from when you started.

THE SENSES

THE FIFTH MONTH

WEEK SEVENTEEN

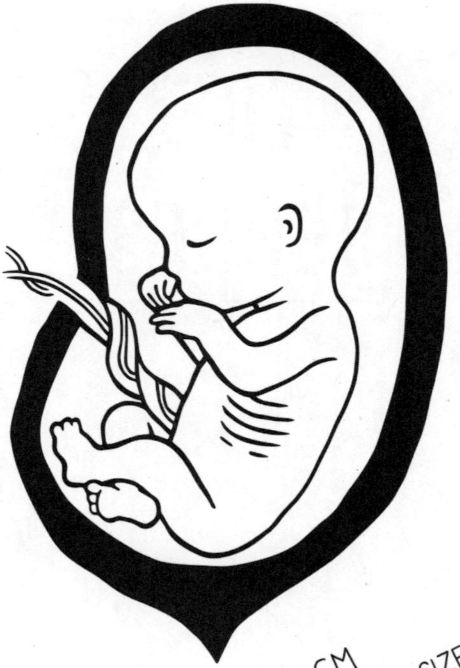

BY THE START of the fifth month you are about as long as a banana. At this point your parents might see you for the first time during an ultrasound scan, and your mum will feel you kicking. With each passing day your muscles get stronger and your skeleton becomes a little harder. Your ears have crept up your neck and are now in the right place. Before long, you will begin reacting to sound. Meanwhile, other senses are already working. The first to come online was your sense of touch. This begins in your second month, when you start reacting to touch around your mouth, and after a while it spreads to other parts of your body. You will often explore your face with your hands, and put your fingers on your lips several times a day.

Taste and smell

YOUR SENSE OF taste most likely begins during month four. By then small taste buds have formed in your mouth, each one a bundle of between fifty and a hundred long, hairy taste cells. The tiny hairs on each taste cell have surface receptors that capture molecules in the food we eat. When the molecule sticks to the receptor, the cell creates a signal telling the brain about what it has found. Each one specialises in a certain type of molecule and its particular taste. Some of them notice when we eat something sour, others if something is sweet, salty, bitter or savoury. New research suggests that there are also cells that react to the taste of fat.

Initially, taste buds are spread all around the mouth, but eventually they end up on different parts of the tongue. However, it is a myth that different areas of the tongue are responsible for each different taste, and that we taste sweetness only with the tip of the tongue, for example. The truth is that there are cells detecting all tastes in every single taste bud. We also have taste buds elsewhere than the tongue – on the palate, for example. Other animals have taste cells all over their bodies. If a fly lands on a slice of apple, it can immediately taste the sweetness of the juice with its feet. The catfish too is covered in so many taste cells that it is almost like a swimming tongue. With this incredible level of sensitivity, it can actually taste small worms hiding under the sand.

When you are in the womb, the taste buds eventually accumulate in small pits on the surface of the tongue and begin to capture molecules from the foetal water you drink. Your sense of taste is nowhere near fully developed yet, however. You might detect a mild sweetness in the foetal water perhaps, but you can't tell that it is salty: you don't gain the ability to recognise that flavour until a few months after your birth. But what goes up must come down: as we enter late middle-age, we lose taste buds, meaning that taste is far more sensitive in children than in older adults.

No matter how many taste buds you have, you will never experience a flavour completely without the help of your nose – as everyone who has eaten a delicious meal while battling a cold will know. If you eat a piece of chocolate while holding your nose you will notice its consistency and its bitterness or sweetness; but the cocoa flavour itself will be missing, because it is the nose that registers this.

Your tiny nose develops sensory cells in month four, but initially both your nostrils are totally closed, blocked by a plug of cells. About a month later these nostril plugs will disappear and you will begin to inhale the foetal water as if it were air. In and out, in and out. While that is happening, molecules from the foetal water will attach to the sensory cells in your nose and a multitude of new impressions will be delivered to your brain. The foetal water, after all, is far more than just salt and water. You are surrounded by a cocktail of substances from both your own and your mother's body. Everything that enters your mother's blood can end up in your underwater home, and that includes flavours from the food she eats. In the United States, courageous researchers sniffed samples of foetal water taken from different pregnant women and could easily tell who had been eating garlic. Other tests have shown that flavours such as mint, aniseed, vanilla and carrot can also be transferred to the womb. Personally I'm convinced that in the womb I often lay bathing in a chocolate pudding and cream flavour; my mother says she ate that constantly while pregnant with me.

Many studies show that we can remember the flavours we experience in the womb. French researchers have found that newborn babies like the smell of aniseed more if the mother has regularly eaten aniseed-flavoured pastilles during pregnancy. In another study, American researchers asked a group of pregnant women to drink a glass of carrot juice four times a week during the final three months of pregnancy. Another group of mothers was asked to avoid carrots completely. About a month after birth, the children were served carrot-flavoured baby food; as you might have guessed, the babies whose mothers drank carrot juice while pregnant enjoyed it far more.

Hearing and balance

THE WOMB IS by no means a quiet environment. When your ears start working you can hear the steady beat of your mother's heart, the roar of her blood and the bubbling of her intestines. Soon these will be joined by many of the noises from outside. Most foetuses start reacting to sounds between weeks twenty and twenty-four, which is at some point during month six. Using ultrasound, researchers have seen that foetuses jump, as if startled, when sounds are played in front of the stomach. The mother's voice is particularly clear to the foetus because it spreads through her entire body. Otherwise, it's the low tones that are most audible, just as you hear bass most clearly when the neighbours are having a party. Many sounds are muted and distorted after passing through skin and muscle. Also, the ears are full of foetal water, which, of course, affects the sound. The consonants and the details get smoothed out, but the rhythms and melodies are just about recognisable. And with every day that passes, your hearing gradually improves.

In order to hear, you need a *cochlea* – a spiral-shaped bone, full of fluid and located at the innermost part of the ear. Inside the cochlea are the hairy sensory cells that tremble to all the tones you hear. The sound creates small waves in the fluid of the cochlea causing the small hairs on the cells to sway, while at the same time sending electrical signals to the brain. Close to the cochlea are three semicircular arches that make up the balance organs. They too are filled with liquid, and when you move your head, you actually create small waves in there, which the sensory cells immediately report to the brain. Together, the three arches register how to move in three dimensions. One of the arches can tell if you are spinning around in a pirouette, another one can tell if you are bending your head forward. Were it not for their frequent reporting to your brain, you would not be able to move about without falling over.

We can tell whether a foetus is affected by sound or touch, but this doesn't tell us if the foetus is *consciously* experiencing its senses. It takes time for the brain's nerve cells to connect properly, allowing them to process these new impressions. What's more, the brain is formed by a combination of inheritance and environment; and everything from chemical substances to experience affects its development. Nerve cells that communicate a lot form stronger relationships with each other. In other words, your senses need to be trained – you get a little better at listening each time your ear picks up a sound. At what point in that process does true conscious experience begin? We don't know.

Hearing problems occur more often in premature babies, and researchers believe this is because the brain simply becomes overwhelmed. By contrast with their former world of soft muted tones in the dark, the child is suddenly met with dazzling lights and the harsh bleeps and pings of hospital equipment. One research group from Harvard University wanted to know whether it was possible to counteract this effect by replicating the womb's environment at the hospital. The lights were dimmed and the babies listened to recordings of their mother's heartbeat and voice. Later, brain scans revealed that these babies had better-developed auditory centres than those who had not received such treatment.

Just as we can remember tastes from the womb, it seems that we are also born with memories of the sounds we heard in there. Newborn babies cry less and breathe more calmly if they can hear a heartbeat. In addition, researchers at Queen's University Belfast have found that nine-month-old foetuses can recognise the title melody of a soap opera if the mother has watched it regularly. By using ultrasound, the researchers saw that the foetus became more active when they played the melody. After they were born it was once again found that the babies became calmer and stopped crying when they heard this familiar tune.

It also seems that we learn to recognise our mother's voice before we are born. When a foetus hears its mother speak, its heart rate increases. Anthony J. DeCasper and William P. Fifer at the University of North Carolina equipped a group of newborn babies with headphones and a specially made dummy. By sucking the dummy quickly or slowly, the babies could choose between a recording of their own mother's voice or the voice of another woman. The mother's voice was the clear favourite. There was only one recording they liked better: one in which the mother's voice had been muffled and distorted to sound more like the voice they had heard in the womb.

The researchers also asked a group of pregnant women to read aloud from a children's book during the final weeks before birth. Each mother sat, twice a day, and read the popular American children's book *The Cat in the Hat* to her unborn child. Then, one day after birth, the baby was equipped with the headphones and special dummy, and once again chose between two recordings. The first was its mother reading *The Cat in the Hat* and the second was its mother reading from another children's book. Once again the verdict was clear: the babies clearly preferred the book that had been read to them while they were still in their mother's tummy.

It is not only people who learn to recognise sounds before they are born. A tiny Australian bird called the superb fairy-wren regularly sings to her eggs while she is brooding, and the hatchlings later use this melody when they scream for food. This is particularly useful against another bird – the artful cuckoo – who likes to lays its own eggs in the same nest. The fairy-wren mother might easily waste food on these unwittingly adopted chicks. But the singing from her own young acts as a secret password, ensuring that only they get the food they need to survive.

Vision

YOU BEGIN RESPONDING to light during the sixth or seventh month. But even though your vision functions, there is not a lot to see in the darkness of the womb.

Only a dim, reddish light penetrates through the clothes, skin, muscles and blood. Nevertheless, you can tell if your mother is lying in the sun: a foetus will turn away when a lamp is shone against its mother's stomach.

The formation of your eyes was already underway back in week four, when your brain was nothing but tiny tubes and your body looked more like a larva. Two hollow stalks grew from either side of the foremost brain tube, both with a tiny sac at the end. A few days later these two sacs rested against the inside of the skin and pressed themselves into two small dishes. That was the beginning of your retinas. But before these cells could begin telling the brain about the light they encountered they needed some help, so they immediately began to send out requests to their neighbours. Some cells started building the lens, which has the task of focusing the light towards the retina. Others built protective materials.

One of the messages these cells use when constructing the eye is made with the help of the gene PAX6. People with a damaged PAX6 gene suffer from a disease called aniridia. A typical characteristic of this is that the coloured part of the eye, the iris, is missing. The eyes of an aniridia patient are neither blue, green nor brown – they consist of just two large, dark pupils in the middle of the eyes. The disease had been known about for over 150 years before the gene causing it was discovered in 1992, but in the years that followed, biologists found that PAX6 was pretty exceptional – and once again it was the fruit fly that took them all by surprise.

Like most insects, the fruit fly is equipped with two extraordinary organs called compound eyes. If you look closely you will see that each one is actually composed of hundreds of tiny red beads, every one of which is an individual eye with its own lens and photosensitive cells. The fruit fly looks in all directions at the same time, and compiles the impressions from every little eye, like a mosaic. To construct them, a gene called Eyeless is used – this gene is also named after what happens when it *doesn't* work (where an unfortunate fruit fly emerges without any eyes at all).

However, when the gene does work, the fruit fly is able to build eyes everywhere. Usually only on the head, but thanks to genetic engineering, researchers have managed to turn on the Eyeless gene in areas where it is usually turned off. The researchers activated the gene in the part of the larva that forms legs and so created a fly with a red eye on each of its six legs. When the gene was turned on in another area, the fruit flies resembled tiny crabs, with eyes protruding from the tips of their antennae.

Once the researchers have mapped the order of the letters in a gene, they can use a computer to search through a database of all known genes and see if it can find any that appear similar. This was precisely what the researchers did with the Eyeless gene, and when the results came up on the screen, they had a surprise. The computer found a match with the human gene PAX6. They had previously found a variant of PAX6 in mice, but that wasn't hugely amazing; mice are mammals like us, and our eyes are quite similar. But the fruit fly was a shock – could it really be possible? Could the same gene used in mammalian eyes really help build the strange red eye-beads of the fruit fly?

In order to investigate the matter further, the researchers decided to cut and paste a little in the fruit fly's genes. This time they tried to give the fruit fly the mouse's variant of PAX6 instead of the Eyeless gene. And the fly cells? They obeyed the mouse's command as faithfully as their own. Both Eyeless and PAX6 work as switches, just like the Hox genes. They switch on other genes that are necessary for building eyes. *Right here we'll have an eye,* decides PAX6, and then other genes take over the actual construction. Even though the command itself is borrowed from a mouse, the fruit fly will then build its usual red insect eyes.

And when this gene commands *your* cells, they will start to build a pair of human eyes.

About two months after conception your eyes are already in position, but it will be some time before you are able to use them. A thin layer of skin will grow over them and keep them closed for almost six months. In any case, the nerve connections to the brain are not yet ready. At this stage your eyes are like two digital cameras without memory cards.

In order to experience seeing, we have to process the visual impression in the cerebral cortex. Someone who is injured in the visual centre in the cerebral cortex will feel completely blind, despite the fact that their eyes actually work. But if you ask that person to reach for an item, they will stretch their hand in the right direction anyway. Even though they feel like they are guessing, they'll do far better than is possible by sheer chance. The reason for this is that there is an additional visual centre deep inside the brain – a souvenir we've inherited from our amphibious forefathers. Just as a frog can instinctively flick out its tongue towards an insect, the blind can seize an object with no idea of what is in front of them.

Even though your eyes are closed for so long in the womb, the cells in your retina need never be bored. After all, the best fun is the fun you make for yourself.

Researchers have measured the activity of retinal nerve cells in various mammals, and have found that they spontaneously send signals to the brain long before vision is developed. In total secrecy, these cells make fake visual impressions behind the sealed eyelids. Well-coordinated waves of electrical activity sweep regularly over the retina, so that the cells sitting beside each other transmit signals to the brain all at once. This helps the nerve cells to connect with each other in the right way.

Despite the head start, your sight is actually your least-developed sense when you are born. At first you are so near-sighted that you are unable to focus on anything more than ten centimetres away. Your eyesight gradually improves, but it still takes a very long while before it comes into its own – as long as fourteen years, by some estimates.

A HAIRY PAST

YOUR PARENTS CANNOT see it on the ultrasound, but something very odd has happened to your body during the fifth month: you have become hairy. Fine white downy hair has grown in a spiral pattern all over your body. You will lose this before you are born, but until then this hair will come in handy: it holds a substance called *vernix caseosa* in place. This white, fatty cream is shed from your skin and works a bit like a moisturiser, protecting your delicate skin from becoming worn or cracked. It also helps you to slide out a little more smoothly when you are born.

In many ways it is really quite strange that we don't make proper fur. All our ape cousins have it, and it provides good protection against both the cold and UV rays from the sun. Instead, we have to make do with short, almost invisible hair on most areas of our bodies. We get sunburned in the summer and cold in the winter. Why would we have got rid of something that seems so useful? Biologists have offered a few suggestions. One possibility is that our ancestors lost their fur as they adapted to life in the warm African savannahs. After leaving the shady forests, it was important to cool down in the baking sun. The solution was to become experts at sweating, a principle as simple as it is brilliant. We have sweat glands all over our skin; when sweat trickles out, it evaporates, drawing heat from the body. While other animals pant when they become too hot, we humans can run very long distances without becoming overheated. We were built for marathon running, and that in itself would have given us a huge advantage when we hunted on the plains. We could keep running until our quarry was on the brink of heat stroke. Sweating and patience were all that was required.

But the naked human encountered a new threat: UV rays from the sun. So our skin developed dark, protective pigments. It wasn't until the first humans moved north from Africa to Europe and Asia that lighter skin types appeared. This light skin gets sunburned faster, but in return uses sunlight more effectively to produce vitamin D.

Heat loss is not the only explanation for human nakedness. Some biologists have suggested that we lost our fur in order to rid ourselves of everything that lived in it. A hairy body is the perfect habitat for ticks, lice and other unpleasant guests who carry dangerous viruses and bacteria. The infection risk is high for social animals living close together. When we learned to make fire, build shelters and fashion clothes, we no longer needed our own fur to keep us warm at night, and the benefits of getting rid of it were greater than the drawbacks. Besides, the winners of evolution are not only the survivors, but those who both survive *and* breed – it's no good living for a hundred years if you never pull! A hairless skin may have been preferable because it sent promising signals about a healthy, parasite-free body – a big hit on the dating scene. An explanation for why we are nevertheless still hairy around the genitals could be that the hair traps the scents and smells that increase sexual attraction.

There is, in any case, one reminder of your furry ancestors that you *have* kept: goose pimples. When you get cold, the muscles around your hair follicles contract automatically, causing the hairs to stand up. For animals with long fur, this is particularly useful since it creates a warm, protective layer around the body and it also makes the animal look bigger and scarier when it feels threatened. But it's a pretty useless reflex for us humans. The pimply skin doesn't make us any warmer, and isn't much use for repelling bears.

FROM WATER TO AIR

THE SIXTH MONTH

WEEK TWENTY-ONE

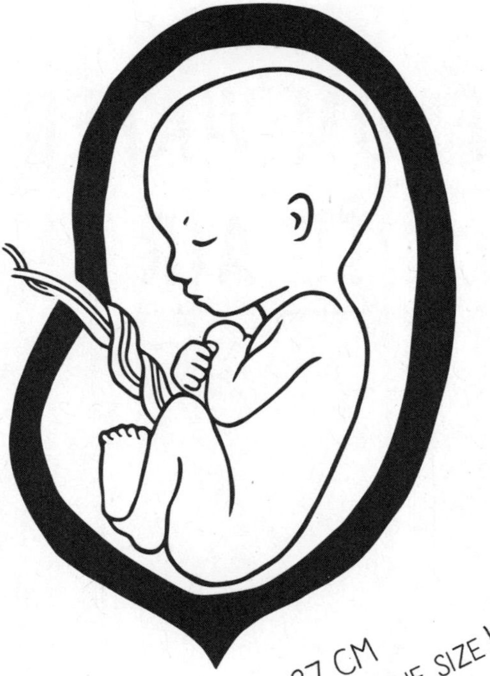

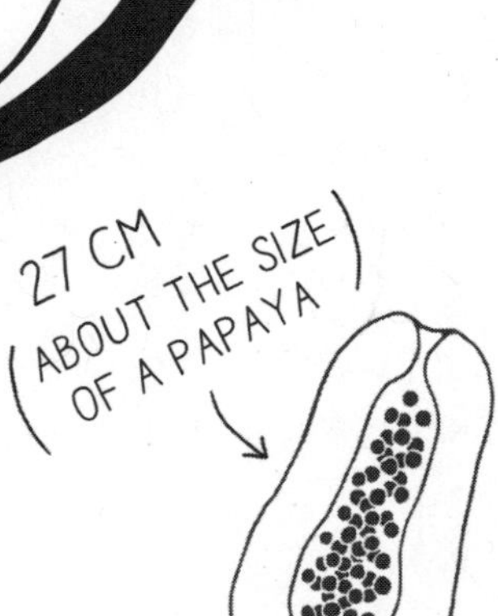

BY THE BEGINNING of the sixth month, you resemble a newborn baby, just smaller and more fragile. Your blood vessels are still visible through your delicate skin, but a fat layer has begun to build up around your skinny body. In the next few weeks your wrinkly skin will smooth out and become less transparent.

If you were born now, you would survive only with help from the hospital. It is possible to save babies born as early as in the twentieth week (pregnancy week twenty-two), but the chances of surviving such an early birth are low. At best, only about one in three make it, and many of them sustain life-lasting damage. By comparison, about 90 per cent survive if the birth occurs five weeks later.

The biggest problem of premature birth is that the lungs are simply not ready to function. They started growing about a month after conception. When you were just a tiny larva a small bud appeared at the top of the intestinal tube. Fish grow a similar bud, even though most of them never develop lungs. Theirs grows into a swim bladder, which allows them to rise or sink without using any muscle power. Some, like the appropriately named lungfish, actually make simple lungs instead of swim bladders. Should the swamp they live in dry up, they can burrow down into a slimy mud-cocoon, and breathe calmly until the rains return.

In your case, the little bud was the beginning of a sophisticated organ that would take many months to complete. The first thing to happen was that the bud sprouted and formed your windpipe. Afterwards, it split into two smaller tubes, and these became your right and left lungs. New tubes sprouted from these like tree branches, and further branches on the branches, until we get to the smallest tubes, which are tipped with a cluster of tiny air sacs. Under a microscope these clusters resemble bunches of grapes. These are your alveoli, and they ensure that the lungs and the blood exchange gases effectively when you start breathing. As you breathe in, the air will spread through the lung branches and inflate the tiny sacs. At the same time, your heart will pump the oxygen-deficient blood into the lungs, where tiny veins thread along the lung branches and coil around the sacs like twine. Since these sacs have extremely thin walls, oxygen molecules can filter directly into the blood and attach themselves to the protein haemoglobin, changing it from a dark, almost black colour to bright red. At the same time, carbon dioxide flows from your blood back into the air sacs – you exhale, and the oxygen-rich blood now flows into your heart's left side, ready to be propelled around the rest of your body.

Du-dunk, du-dunk. Breathe in, breathe out, and so on, from birth until death. To ensure that the blood isn't caught in a never-ending loop through your lungs, there is a solid wall of muscle separating the heart's left and right sides. But obviously, this whole mechanism only makes sense once you are getting your oxygen from your lungs. While you remain in the womb, you aren't, and so your blood has to go on an entirely different journey to collect oxygen: through the umbilical cord and into the placenta. As soon as your blood has enough oxygen it returns to your body and enters the right side of your heart. From there it should carry on to the lungs, but because there is nothing to be collected from the water-filled sacs just yet, it takes a shortcut, through a tiny hole and straight into the heart's left side.

When you draw breath for the very first time, a small valve will close over this hole, sealing it for ever; and instantly sending your blood on its new course, through the lungs – something it will faithfully continue doing for the rest of your life. But if this closure does not go as it should, then you will continue to have a small hole in your heart after you are born. This is one of the most common congenital heart defects. Fortunately, the hole usually closes up by itself after a while. If it doesn't, then with each heartbeat a small volume of blood will flow from the left side of the heart to the right side. This blood gets pumped an additional round through the lungs and gives the heart unnecessary extra work. To avoid the heart being overloaded, large holes can be closed with heart surgery.

Speaking of holes in the heart, I have yet another little fruit fly story to tell you. Believe it or not, there is, in fact, a simple, tube-like heart concealed within the fruit fly's tiny body. Flies don't have blood or veins, but the little tube pulsates and moves all the fluid that surrounds the insect's organs. In the 1980s the American researcher Rolf Bodmer searched for the gene that controlled the development of the fruit fly's nervous system, but instead he stumbled upon a gene that proved to play a decisive role in the development of this little heart tube. If this gene was destroyed, then the fruit fly would literally become heartless. Bodmer therefore chose to call the gene Tinman, after the character in *The Wizard of Oz*.

A few years later, another research group studied the genes of patients who had been operated on to correct congenital holes in the heart, and they found that all of them had mutations within the same area on chromosome number five. Hadn't they seen a similar DNA code somewhere before? As it turned out, humans also have a variant of the Tinman gene (which was, sadly, given the far more boring name Nkx2.5). Humans and fruit flies have their own versions of the same gene, which they use to build completely different hearts. Once again, evolution has preserved an ancient innovation.

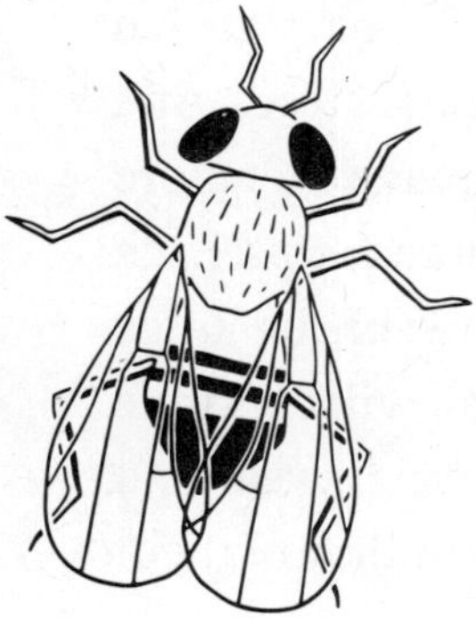

On the other hand, if it is lungs you are interested in, then the fruit fly isn't much use as a test animal. Like other insects, the fruit fly acquires oxygen in a completely different way from us. If you study an insect with a magnifying glass, you will see tiny openings along its body. These allow air in, which is then distributed round its body through a network of tiny pipes. It's a simple arrangement that works outstandingly over short distances, but is useless for large organisms like us: there just wouldn't be enough oxygen to reach our innermost cells. About 300 million years ago there were dragonflies as big as seagulls, but this was possible only because the air was far richer in oxygen back then. Nowadays it is really quite essential for insects to stay small. So you can breathe easy: giant insects will not be making a comeback.

We humans are dependent on having a pair of well-functioning lungs from the very moment the umbilical cord is cut. Two months after conception, you were already practising breathing by inhaling foetal water in and out of your unfinished lungs; your chest rose and sank rhythmically, as if you actually were breathing. By the sixth month, your lungs have grown out into large, multi-branched tree-like structures. And your cells are working persistently to make ever more alveoli; the tiny air sacs that sit at the branch tips. Each new alveolus increases the surface area of the lungs, thereby increasing oxygen uptake. You will continue to make new air sacs until you are about eight years old, by which time you will have about 300 million of them.

In the final months, your lungs have begun tackling another vital task: the production of a substance called surfactant. This prevents your lungs from sticking together when you exhale, and without enough of it there is a risk of your lungs collapsing. When researchers discovered how to produce artificial surfactant in the 1980s, survival rates for premature babies improved significantly. Doctors can now spray artificial surfactant straight into the lungs of children who are born prematurely. In addition, premature babies are placed in an incubator – a sealed intensive-care bed with transparent walls – where it is possible to adjust temperature, humidity and oxygen levels. If necessary, the doctors will connect a respirator that blows air into the lungs.

Today's technology makes it possible to save babies that would otherwise have died. Will it ever be possible to move an entire pregnancy out of the womb? In 2016 two research groups succeeded, for the first time, in growing human embryos in a laboratory for more than a week. The scientists saw how the cell vesicle attached itself to the laboratory dish and then watched it continue developing for a fortnight. It is possible they could have kept the embryo alive even longer, but at this point the experiment was stopped for ethical and legal reasons.

At the Children's Hospital of Philadelphia, researchers have recently tested an artificial womb on prematurely born lambs. The system consists of a transparent plastic bag filled with synthetic foetal water and a machine that delivers oxygen and nutrition through an 'umbilical cord'. Inside the plastic bag, the lamb is able to swallow and breathe the 'foetal water', enabling its lungs to develop normally. Scientists hope this technology can be used for people in the future, but stress that it is only intended to assist the difficult transition from water to air. The simple salt solution and machine-driven placenta can in no way replace the complex environment that exists in a human body.

It is not just the lungs that need nine months in the womb. The final months are also important for the development of the brain. During the seventh month, the brain reaches a milestone. Its electrical activity can now be measured as synchronised, regular waves. Previously, there have only been random flashes of brain activity, but by measuring these brainwaves, researchers have found that the foetus is mostly asleep while in the womb. The low oxygen level, together with tranquillising substances from the placenta, ensures that the foetus is awake for less than 10 per cent of the day. Otherwise it alternates between calm and active sleeping phases. During the active phases, also known as REM sleep, the eyes move rapidly from side to side behind closed eyelids. The brain too appears to wake up momentarily in the sleeping body: the large calm brainwaves turn quick and short, typical of someone who is awake.

REM sleep is found in most mammals and birds, but researchers still don't agree on its true purpose. Adult humans have several periods of REM sleep each night, and it's also the period when we usually dream the most. Rats seem to have the ability to dream too – about finding their way to a tasty treat, for example. Researchers at the Massachusetts Institute of Technology measured the brain activity of rats as they navigated through a maze in search of chocolate. Afterwards, they studied the rats while they slept, and saw that the same parts of the brain were firing signals. Similar observations were made of zebra finches, except it turns out they dream about singing, according to researchers at the University of Chicago. When these birds tweet a melody, they fire specific nerve cells for each tone. When the researchers monitored the birds as they were sleeping, they saw that the same nerve cells became active again – as if the birds were practising the song in their dreams.

Do foetuses dream? No one knows for sure, but we do know that foetuses and babies spend far more of their time in REM sleep than adults. While the REM phases account for less than a quarter of the sleep in an adult human, they account for more than half of a foetus's sleep. Why might that be? Recent studies on mice have shown that a kind of clean-up operation takes place in the brain during REM sleep, where unnecessary connections between nerve cells are removed. That invites the suspicion that this sleep phase is important for brain development. Could that also be what adults are doing when they dream at night? After all, our brains will never be the finished article – they will continue to change as long as we live, learn and remember.

Whether you are awake or asleep, your body carries on preparing for life outside the womb. From the seventh month onwards, your appearance doesn't change much more, except that you become chubbier. Small folds appear on your arms and legs, and in recent weeks, your weight has increased by an average of 14 grams a day.

Foetuses put on both regular fat and something called brown fat. The cells in brown fat are experts at producing heat, and will be useful in the cold world outside the womb. Adults are far better equipped to cope with cold than babies; they have more muscle, are better at trembling and can move somewhere warmer of their own free will. That's why we tend to lose our brown fat as we grow older. Bears, however, produce masses of brown fat every summer, which they use to keep warm as they hibernate through the long, cold winter.

During the final weeks, things start getting a bit cramped in the womb. The days when you could tumble and roll about in there are over. Towards the end you are lying in what, for very good reasons, is called the foetal position – with your long legs folded up against your chest. You've got just about enough space to kick firmly at your mother's ribs and tummy. If you are lying in the same position as most other babies, then your head will soon press against the birth canal.

And then it happens.

THE END — OR THE BEGINNING

FULL TERM

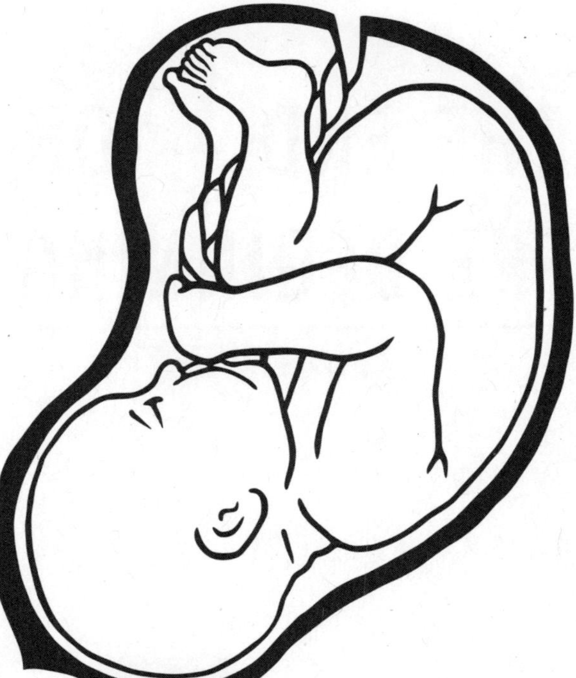

36 CM
OR 50 CM INCLUDING THE LEGS

(ABOUT THE SIZE OF A NEWBORN BABY)

The End

A KANGAROO BIRTH GOES almost completely unnoticed. It happens silently, only a month after conception. From between the kangaroo mother's legs, a worm-like creature the size of a jellybean crawls out. Its skin is smooth and transparent, and red veins run through its little body. The newborn's hind legs have not yet grown properly, but it clings to its mother's thick pelt with its front legs and pushes itself upwards. The mother leans forward, licking her fur to create a path for the baby, and slowly the tiny creature crawls up to her pouch. Here it will spend the next nine months, growing in its own safe space, with direct access to its mother's milk – until finally one day it hops out, ready to explore the world.

The spotted hyena has it much worse. It must push out its cubs through a penis-like tube that barely accommodates the larger ones. Often the tube splits during birth, causing the death rate for first-time hyena mothers to be sky-high. You would have thought the species would have made themselves extinct by now, but somehow they seem to manage quite well. The cubs at least get plenty of time to grow and mature in the womb, and are well prepared for the brutal reality waiting for them. Shortly after birth, the cubs are ready to kill, with powerful jaws and fully grown teeth.

You can't really say the same about newborn humans. When you are born you will be about as helpless as the little kangaroo joey. While most animals are perfectly mobile shortly after birth, you will be incapable of doing anything besides sucking milk, sleeping and screaming. However, you will develop an impressively large brain. Had your skull not been made of movable bone-plates, there would be no room for the remarkable organ that is growing in there.

It doesn't help much that, between five and seven million years ago, our ancestors began to walk on two legs. This weird method of getting around has affected the shape of our skeletons. If you compare a chimpanzee with a human, you will see clear differences in the bones that connect the legs to the back. Our pelvis is shorter, wider and more bowl-shaped than the chimpanzee's. This design supports the spine and allows us to move quickly and efficiently on two legs. It also ensures that we can carry the weight of all our internal organs, in addition to a growing foetus if necessary. Things are different for chimpanzees. Unlike us, they support the weight of their organs with the abdominal muscles, so that the weight is evenly distributed over a large area, like in a hammock. We have to carry more than half our body weight with the muscles and bones in the pelvis. If the distance between the bones ever became too large, our organs might fall out.

Like most mammals, human births pass through the pelvis. Even if the cervix and the vagina are expandable, the pelvic bones place rigid limitations on how wide the birth canal can be. A number of monkey species have the same problem – their heads are so large, they barely fit through the pelvic opening. Birth complications among both species are not uncommon. Because the human pelvis is adapted for two-legged walking, however, things become even more complicated. The pelvic opening is bigger at the front than at the back, which means you have to rotate completely during the birth, so that your back points towards your mother's stomach.

Thankfully, you don't necessarily have to do that part alone. One big advantage that we humans have relative to most other species is the partners, midwives and other nursing staff who help us during birth. With monkeys it is most common for the mother to give birth in solitude. There are some interesting exceptions: in the lush rainforests of Central and South America, for example, there is a monkey species in which the fathers lick and hold the baby right after the birth. In another species, experienced female monkeys have been observed helping first-time mothers by pulling the baby out with their hands, just like a midwife.

One thing monkeys do have going for them is that the baby often acts as its own midwife, hauling itself forward with its arms. Once it gets the chance, the baby monkey grabs its mother's fur and climbs up to her breasts. Human babies are incapable of doing the same – our motor skills are far too poor. That said, when my father read this section, he recalled that I was quite active during my own birth. Just after my head emerged, I squeezed my elbows out and pushed myself forward. (I was obviously in a hurry to see the world.)

Either way, evolution really should have fixed human birth a long time ago. Squeezing out helpless babies through a far-too-narrow opening doesn't seem like the wisest strategy for survival. Most mammals' brains are roughly half their full size before birth. We humans, on the other hand, only manage to grow ours to a third of its full size. Our brains continue to mature at almost the rate of a foetal brain after we have been born. In the first few months, our brains make huge numbers of new connections, doubling their size in just one year. Maybe that's why human infants are so helpless: our brains are simply too under-developed at birth. But this low level of development can also be a huge advantage: it makes us adaptable and eager to learn. Once we are outside the womb we can shape our brains according to our surroundings and experiences.

No matter how helpless you still are, though, at some point you'll just have to get out. If your head gets bigger, it will be impossible to squeeze out through the narrow birth canal. In addition, your brain and your growing body need ever more energy, and now that the end of the pregnancy is approaching, your mother is struggling to provide it. The brain is a demanding organ. It is so ravenous that it consumes about a fifth of all the energy you receive. If you ate five spoons of porridge, the last spoon would go to your brain alone. Access to oxygen is also quite poor in the womb – you receive only a fifth of what you can get outside – so if your brain is to keep growing, then your lungs must get enough fresh air. You simply have to get out, you have to breathe. Right now.

The Beginning

WAS IT REALLY you who decided what day you would leave your home of nine months, or were you just evicted by your mother? The truth is, probably a bit of both. Researchers still haven't formed a complete picture of the process that initiates childbirth. It is a secret conversation between your mother's cells, the skin that surrounds you in the womb, the placenta and your own cells, and it begins several weeks before the event. In 2015 American researchers found that mouse foetuses create a signal in their lungs that helps to start the birth. It is possible that something similar happens in humans. Perhaps your lungs whispered one of the first messages: *Hello, brain, we're ready to breathe. Soon you'll get all the oxygen you need.*

But it is not enough to have two breath-ready lungs. In the 1950s sheep farmers in Idaho were afflicted by a very unpleasant event: about a quarter of their lambs were born with abnormalities. Their brains were deformed, and they each had just a single eye in the middle of their heads. In addition, the malformed lambs were born extremely late. A sheep is usually pregnant for about 150 days, but these lambs went over 200 days before being born. There were also a number of sheep that were unable to start giving birth, and the lambs had to be delivered by Caesarean section.

What had gone wrong? The culprit was found to be a poisonous white lily that grew where the sheep grazed, although the precise mechanism was unclear. Several decades later, in the 1990s, however, researchers found that the poison in this flower prevents cells from listening to a protein message that is sent early in embryonic development. This leads to several malformations; one of which occurred because the cells didn't hear the command to divide the foremost brain tube in two. But why would this actually delay the birth? Veterinarians had previously reported lengthy pregnancies in cows bearing calves with severe brain malformations. Similar cases have also been reported in humans. And these clues pointed at the same thing: the foetus's brain is involved with initiating the birth.

When the time has come, the nerve cells in your brain speak to the hormone glands: *Get ready, it's happening!* Then, among other things, your hormone glands transmit this message to the rest of your body with an increase of cortisol. Soon a chemical call-out is booming through your blood, and your cells begin to prepare. Tiny pumps in your lungs begin to remove fluid and the cells begin producing more surfactant. While this is happening, the fat cells break down more fat to get energy. When the hormone reaches the placenta, it affects your mother as well. Since the very start, her body has told itself *Not yet*. Calming signals have flowed from the placenta and have prevented the womb's muscles from contracting strongly. For months, these muscles have been sat waiting, just contracting gently every now and then – but when the cortisol flows into the womb, it triggers a chain reaction: hormone signals change from *Not yet*, to *Let's go*; The contractions become stronger and more frequent; at the same time, the muscles create more receptors to receive the hormones – almost as though they prick up their ears and start listening out for signals. And as soon as your head presses against the nerves in the birth canal, it causes your mother to release even more hormones. Her muscle cells respond by contracting, rhythmically and regularly, and those contractions become increasingly powerful. Little by little, they push your tiny body forward.

Then, your safe dark pool bursts. Your head is clenched firmly. The contractions, pressing on the placenta and the umbilical cord, cut off your oxygen supply periodically. It's almost as though you are being suffocated. Your body responds by releasing unusually large amounts of the so-called stress hormones adrenaline and noradrenaline. Later in life you'll create these hormones whenever you face danger. They ensure that your blood pressure increases and that your heart beats faster. The cells rapidly break down your energy reserves. Blood is diverted from your skin and intestines, to what are now the most important parts of your body: heart, brain and muscles. You flex your muscles and prepare yourself for two possible alternatives: *fight or flight*.

During your own birth, your stress hormone level is at the highest it will ever be. Even though your mother is probably quite stressed herself, it doesn't compare to your hormone rush. In fact, even a heart attack can't trigger the same reaction. But though it perhaps doesn't sound like it, these stress hormones are very good for you: they help you tackle the stress you experience on your way out of the womb and prepare your body for life outside. For example, stress hormones make the cells break down the nutrition you'll need to live on when you lose access to the placenta. They also make sure that your lungs clear themselves of fluid, so that you are ready to take your first breath.

Which is any moment now. Soon two unfamiliar hands will take hold of your head, the dazzling light will hit your eyes, and your lungs will fill with air for the very first time.

You will breathe.

And what happens next? You know more about that than me.

REFERENCES

Images

SIZE REFERENCE FROM week 1 is from: Nesheim, Britt-Ingjerd, Foster, in *Store medisinske leksikon*; https://sml.snl.no/foster; others are taken from Moore, K. L., Persaud, T. V. N. and Torchia, M. G. (2016), *The Developing Human: Clinically Oriented Embryology* (10th edn). Philadelphia, PA: Saunders Elsevier. See the tables on pages 76 and 92.

Textbooks used in all chapters

I HAVE USED many standard textbooks on embryology, developmental biology and cell biology as sources. The two most important are Moore, K. L., Persaud, T. V. N. and Torchia, M. G. (2016), *The Developing Human: Clinically Oriented Embryology* (10th edn), Philadelphia, PA: Saunders Elsevier, and Gilbert, S. F. (2010), *Developmental Biology* (9th edn), Sunderland, MA: Sinauer Associates. I strongly recommend them if you want to dive a bit deeper into the story of the making of you.

Below is an overview of other sources that I've used, sorted by chapter.

The Race

Bahat, A., Caplan, S. R. and Eisenbach, M. (2012), Thermotaxis of human sperm cells in extraordinarily shallow temperature gradients over a wide range, PLOS ONE, 7(7): e41915

Eisenbach, M. and Giojalas, L. C. (2006), Sperm guidance in mammals – an unpaved road to the egg, *Nature Reviews Molecular Cell Biology*, 7(4): 276–85

van der Ven, H. H., Al-Hasani, S., Diedrich, K., Hamerich, U., Lehmann, F. and Krebs, D. (1985), Polyspermy in in vitro fertilization of human oocytes: frequency and possible causes. *Annals of the New York Academy of Sciences*, 442: 88–95

The Hidden Universe

Clift, D. and Schuh, M. (2013), Restarting life: fertilization and the transition from meiosis to mitosis, *Nature Reviews Molecular Cell Biology*, 14(9): 549–62; doi: 10.1038/nrm3643

Gilbert, S. F. and Barresi, J. F. (2016), *Developmental Biology* (11th edn), Sunderland, MA: Sinauer Associates. Additional article, Chapter 7: 'Anton van Leeuwenhoek and his perception of spermatozoa'; http://11e.devbio.com/wt070102.html

Gjersvik P. (2008), Sædcellen, *Tidsskrift for Den norske legeforening*, 3: 128–265

Harris, H. (2002), *Things Come to Life: Spontaneous Generation Revisited*, Oxford: Oxford University Press

Lawrence, C. R. (2008), Preformationism in the Enlightenment, *Embryo Project Encyclopedia*; http://embryo.asu.edu/handle/10776/1926

Leeuwenhoek, A. van (1677) Letter no. 35, to William Brouncker, November 1677. The whole letter can be read in Dutch and in English translation on the web site of DBNL – De Digitale Bibliotheek voor de Nederlandse Letteren: www.dbnl.org/

Maienschein, J. (2005), Epigenesis and preformationism, *Stanford Encyclopedia of Philosophy*; http://plato.stanford.edu/entries/epigenesis/

Pasteur, L. (1864), On spontaneous generation. An address delivered by Louis Pasteur at the Sorbonne Scientific Soirée of 7 April 1864

The Recipe for a Human

Dahm, R. (2005), Friedrich Miescher and the discovery of DNA, *Developmental Biology*, 278(2): 274–88; doi: 10.1016/j.ydbio.2004.11.028

O'Connor, C. (2008), Isolating hereditary material: Frederick Griffith, Oswald Avery, Alfred Hershey, and Martha Chase, *Nature Education*, 1(1): 105

Pray, L. (2008), Discovery of DNA structure and function: Watson and Crick, *Nature Education*, 1(1): 100

The Invasion

Bayes-Genis, A., Bellosillo, B., de la Calle, O., Salido, M., Roura, S., Ristol, F. S. and Cinca, J. (2005), Identification of male cardiomyocytes of extracardiac origin in the hearts of women with male progeny: male fetal cell microchimerism of the heart, *Journal of Heart and Lung Transplantation*, 24(12): 2179–83; doi: 10.1016/j.healun.2005.06.003

Bianconi, E., Piovesan, A., Facchin, F., Beraudi, A., Casadei, R., Frabetti, F. and Canaider, S. (2013), An estimation of the number of cells in the human body, *Annals of Human Biology*, 40(6): 463–71; doi: 10.3109/03014460.2013.807878

Brosens, J. J., Salker, M. S., Teklenburg, G., Nautiyal, J., Salter, S., Lucas, E. S. and Macklon, N. S. (2014), Uterine selection of human embryos at implantation, *Scientific Reports*, 4: 3894; doi: 10.1038/srep03894

Chan, W. F., Gurnot, C., Montine, T. J., Sonnen, J. A., Guthrie, K. A. and Nelson, J. L. (2012), Male microchimerism in the human female brain, PLOS ONE, 7(9): e45592; doi: 10.1371/journal.pone.0045592

Felker, G. M., Thompson, R. E., Hare, J. M., Hruban, R. H., Clemetson, D. E., Howard, D. L. and Kasper, E. K. (2000), Underlying causes and long-term survival in patients with initially unexplained cardiomyopathy, *New England Journal of Medicine*, 342(15): 1077–84; doi: 10.1056/nejm200004133421502

Gellersen, B. and Brosens, J. J. (2014), Cyclic decidualization of the human endometrium in reproductive health and failure, *Endocrine Reviews*, 35(6): 851–905; doi: 10.1210/er.2014-1045

Kara, R. J., Bolli, P., Karakikes, I., Matsunaga, I., Tripodi, J., Tanweer, O. and Chaudhry, H. W. (2012), Fetal cells traffic to injured maternal myocardium and undergo cardiac differentiation, *Circulation Research*, 110(1): 82–93; doi: 10.1161/circresaha.111.249037

Melford, S. E., Taylor, A. H. and Konje, J. C. (2014), Of mice and (wo) men: factors influencing successful implantation including endocannabinoids, *Human Reproduction Update*, 20(3): 415–28; doi: 10.1093/humupd/dmt060

National Institutes of Health (NIH) History (2003), A timeline of pregnancy testing; https://history.nih.gov/exhibits/thinblueline/timeline.html

Oron, E. and Ivanova, N. (2012), Cell fate regulation in early mammalian development, *Physical Biology*, 9(4): 045002; doi: 10.1088/1478-3975/9/4/045002

Teklenburg, G., Salker, M., Molokhia, M., Lavery, S., Trew, G., Aojanepong, T. and Macklon, N. S. (2010), Natural selection of human embryos: decidualizing endometrial stromal cells serve as sensors of embryo quality upon implantation, PLOS ONE, 5(4): e10258; doi: 10.1371/journal.pone.0010258

Wang, Y. and Zhao, S. (2010), *Vascular Biology of the Placenta*, San Rafael, CA: Morgan & Claypool Life Sciences

Natural Clones and Unknown Twins

Davies, J. A. (2014), *Life Unfolding: How the Human Body Creates Itself*, Oxford: Oxford University Press

Friedman, L. F. (2014), The stranger-than-fiction story of a woman who was her own twin, *Business Insider*, 2 February; http://uk.business insider.com/lydia-fairchild-is-her-own-twin-2014-2/

Kean, S. (2013), The you in me, *Psychology Today*, 11 March; https://www.psychologytoday.com/articles/201303/the-you-in-me/

Kramer, P. and Bressan, P. (2015), Humans as superorganisms, *Perspectives on Psychological Science*, 10(4): 464–81; doi: 10.1177/1745691615583131

Milo, R. and Phillips, R. (2015), *Cell Biology by the Numbers*, New York: Garland Science; http://book.bionumbers.org/how-many-genes-are-in-a-genome/

National Human Genome Research Institute (2016), An overview of the Human Genome Project, 11 May; https://www.genome.gov/12011238/an-overview-of-the-human-genome-project/

National Human Genome Research Institute. (2016). The cost of sequencing a human genome, 6 July; https://www. genome.gov/sequencingcosts/

O'Shea, K. (2014). Medical mystery: woman gives birth to children, discovers her twin is actually the biological mother, *Philly.com*, 4 February; http://www.philly.com/philly/health/science/Medical_mystery_Woman_gives_birth_to_children_discovers_her_twin_is_actually_the_biological_mother.html

Robson, D. (2015). Is another human living inside you? BBC *Future*, 18 September; http://www.bbc.com/future/story/20150917/is-another-human-living-inside-you

Tao, X., Chen, X., Yang, X. and Tian, J. (2012), Fingerprint recognition with identical twin fingerprints, PLOS ONE, 7(4): e35704; doi: 10.1371/journal.pone.0035704

van Dijk, B. A., Boomsma, D. I. and de Man, A. J. (1996), Blood group chimerism in human multiple births is not rare, *American Journal of Medical Genetics*, 61(3): 264–8; doi: 10.1002/(SICI)1096-8628(19960122)61:3<264::AID-AJMG11>3.0.CO;2-R

The Contours of a Body

Brown, Paul (1999), Listening to the heart of the ocean, *The Guardian*, 29 July; https://www.theguardian.com/science/1999/jul/29/technology

Fielder, S. E. (2016), Resting heart rates, in *Merck Veterinary Manual*, Kenilworth, NJ: Merck & Co.

Hodge, R. (2010), *Developmental Biology: From a Cell to an Organism*, New York: Facts on File

Levine, H. J. (1997), Rest heart rate and life expectancy, *Journal of the American College of Cardiology*, 30(4): 1104–6.

Nesheim, Britt-Ingjerd (2014), Foster, *Store medisinske leksikon*, 6 November; https://sml.snl.no/foster

Cell Language for Beginners

Ahmed, A. M. (2002), History of diabetes mellitus, *Saudi Medical Journal*, 23(4): 373–8

Eknoyan, G. and Nagy, J. (2005), A history of diabetes mellitus or how a disease of the kidneys evolved into a kidney disease, *Advances in Chronic Kidney Disease*, 12(2): 223–9; doi: 10.1053/j.ackd.2005.01.002

Vaaler, Stein and Berg, Jens Petter (2016), Diabetes, *Store medisinske leksikon*, 11 August; https://sml.snl.no/diabetes

The Art of Building a Fruit Fly

Carroll, S. B. (2005), *Endless Forms Most Beautiful: The New Science of Evo Devo and the Making of the Animal Kingdom*, New York: Norton

Gehring, W. J. (1998), *Master Control Genes in Development and Evolution: The Homeobox Story*, New Haven: Yale University Press

Jacob, F. and Monod, J. (1961), Genetic regulatory mechanisms in the synthesis of proteins, *Journal of Molecular Biology*, 3(3): 318–56; doi: 10.1016/S0022-2836(61)80072-7

Jacobson, Brad (2010), Homeobox genes and the Homeobox, *Embryo Project Encyclopedia*, 11 October; http://embryo.asu.edu/handle/10776/2070

Laughon, A. and Scott, M. P. (1984), Sequence of a Drosophila segmentation gene: protein structure homology with DNA-binding proteins, *Nature*, 310: 25; doi: 10.1038/310025a0

Lewis, E. B. (1978), A gene complex controlling segmentation in Drosophila, *Nature*, 276: 565; doi: 10.1038/276565a0

McGinnis, W., Garber, R. L., Wirz, J., Kuroiwa, A. and Gehring, W. J. (1984), A homologous protein-coding sequence in Drosophila homeotic genes and its conservation in other metazoans, *Cell*, 37(2): 403–8; doi: 10.1016/00928674(84)90370-2

McGinnis, W., Levine, M. S., Hafen, E., Kuroiwa, A. and Gehring, W. J. (1984), A conserved DNA sequence in homoeotic genes of the Drosophila Antennapedia and bithorax complexes, *Nature*, 308: 428; doi: 10.1038/308428a0

Myers, P. (2008), Hox genes in development: the Hox code, *Nature Education*, 1(1): 2

Nüsslein-Volhard, C. (2006), *Coming to Life: How Genes Drive Development*, San Diego, CA: Kales Press

Wolpert, L. (1991), *The Triumph of the Embryo*, Oxford: Oxford University Press

An Heirloom from the Ocean

Brooker, R. J. (2011), The origin and history of life, *Biology* (2nd edn), New York: McGraw-Hill, s. 438–58

Darwin, C. and Johansen, K. (2005), *On the Origin of Species by Means of Natural Selection, or the Preservation of Favoured Races in the Struggle for Life*, Oslo: Bokklubben.

Shubin, N. (2009), *Your Inner Fish: The Amazing Discovery of Our 375-Million-Year-Old Ancestor*, London: Penguin

Fleshing Out the Plan

Evensen, Stein A. and Wislø, Finn (2017), Blod, *Store medisinske leksikon*, 1 March; https://sml.snl.no/blod

Kretzschmar, D., Hasan, G., Sharma, S., Heisenberg, M. and Benzer, S. (1997), The Swiss Cheese mutant causes glial hyperwrapping and brain degeneration in Drosophila, *Journal of Neuroscience*, 17(19): 7425–32.

Lukacsovich, T., Yuge, K., Awano, W., Asztalos, Z., Kondo, S., Juni, N. and Yamamoto, D. (2003), The Ken and Barbie gene encoding a putative transcription factor with a BTB domain and three zinc finger motifs functions in terminalia development of Drosophila, *Archives of Insect Biochemistry and Physiology*, 54(2): 77–94; doi: 10.1002/arch.10105

NASA Education (2004), Bones in space, 19 August; www.nasa.gov/audience/foreducators/postsecondary/features/F_Bones_in_Space.html

NASA Science (2001), Space bones, 1 October; http://science.nasa.gov/science-news/science-at-nasa/2001/ast01oct_1/

Office of the Surgeon General (US) (2004), The basics of bone in health and disease, *Bone Health and Osteoporosis: A Report of the Surgeon General, Rockville, MD, US Dept of Health and Human Services*; https://www.ncbi.nlm.nih.gov/books/NBK45504/

Tickle, C. and Towers, M. (2017), Sonic Hedgehog signaling in limb development, *Front Cell Developmental Biology*, 5: 14; doi: 10.3389/fcell.2017.00014

Tran, V. (2017), *Muskel-og skjelettsystemets utvikling, Norsk Helseinformatikk* 2017, 3 April; https://nhi.no/familie/graviditet/svangerskap-og-fodsel/fosterutvikling/muskel-og-skjelettsystemets-utvikling/

Varjosalo, M. and Taipale, J. (2008), Hedgehog: functions and mechanisms. *Genes & Development*, 22(18): 2454–72; doi: 10.1101/gad.1693608

Sex and Sea Worms

Berec, L., Schembri, P. J. and Boukal, D. S. (2005), Sex determination in *Bonellia viridis (Echiura: Bonelliidae)*: population dynamics and evolution, *Oikos*, 108(3): 473–84

Gallup, G. G. Jr, Finn, M. M. and Sammis, B. (2009), On the origin of descended scrotal testicles: the activation hypothesis, *Evolutionary Psychology*, 7(4): 517–26; doi: 10.1177/147470490900700402

Jost, A., Vigier, B., Prepin, J. and Perchellet, J. P. (1973), Studies on sex differentiation in mammals, *Recent Progress in Hormone Research*, 29: 1–41

US National Library of Medicine, Genetics Home Reference (2010), Y chromosome; https://ghr.nlm.nih.gov/chromosome/Y

Warner, R. R. and Swearer, S. E. (1991), Social control of sex change in the bluehead wrasse, Thalassoma bifasciatum (Pisces: Labridae), *The Biological Bulletin*, 181(2): 199–204; doi: 10.2307/1542090

Willard, H. F. (2003), Tales of the Y chromosome, *Nature*, 423: 810–11, 813; doi: 10.1038/423810a

Wilson, C. A. and Davies, D. C. (2007), The control of sexual differentiation of the reproductive system and brain, *Reproduction*, 133(2): 331–59; doi: 10.1530/rep-06-0078

Secret Preparations

Holck, P. (2017), Nyre, *Store medisinske leksikon*, 27 September; https://sml.snl.no/nyre

Saint-Faust, M., Boubred, F. and Simeoni, U. (2014), Renal development and neonatal adaptation, *American Journal of Perinatology*, 31: 773–80; doi: 10.1055/s-0033-1361831

The Brain's Inner Wanderings

Hepper, P. G., Shahidullah, S. and White, R. (1991), Handedness in the human foetus, *Neuropsychologia*, 29(11): 1107–11

Hepper, P. G., Wells, D. L. and Lynch, C. (2005), Prenatal thumb sucking is related to postnatal handedness, *Neuropsychologia*, 43(3): 313–15; doi: 10.1016/j.neuropsychologia.2004.08.009

Lagercrantz, H. and Ringstedt, T. (2001), Organization of the neuronal circuits in the central nervous system during development, *Acta Paediatrica*, 90(7): 707–15

Linden, D. J. (2007), *The Accidental Mind*, Cambridge, MA: The Belknap Press of Harvard University Press

Stiles, J. and Jernigan, T. L. (2010), The basics of brain development, *Neuropsychology Review*, 20(4): 327–48; doi: 10.1007/s11065-10-91484

Xie, L., Kang, H., Xu, Q., Chen, M. J., Liao, Y., Thiyagarajan, M. and Nedergaard, M. (2013), Sleep drives metabolite clearance from the adult brain, *Science*, 342: 373

The Senses

Besnard, P., Passilly-Degrace, P. and Khan, N. A. (2016), Taste of fat: a sixth taste modality? *Physiological Reviews*, 96(1): 151–76; doi: 10.1152/physrev.00002.2015

Colombelli-Négrel, D., Hauber, Mark E., Robertson, J., Sulloway, Frank J., Hoi, H., Griggio, M. and Kleindorfer, S. (2012), Embryonic learning of vocal passwords in superb fairy-wrens reveals intruder cuckoo nestlings, *Current Biology*, 22(22): 2155–60; doi: 10.1016/j.cub.2012.09.025

DeCasper, A. J. and Fifer, W. P. (1980), Of human bonding: newborns prefer their mothers' voices, *Science*, 208: 1174–6

DeCasper, A. J. and Spence, M. J. (1986), Prenatal maternal speech influences newborns' perception of speech sounds, *Infant Behavior and Development*, 9(2): 133–50; doi: 10.1016/0163-6383(86)90025-1

Graven, S. N. and Browne, J. V. (2008), Auditory development in the foetus and infant, *Newborn and Infant Nursing Reviews*, 8(4): 187–93; doi: 10.1053/j.nainr.2008.10.010

Halder, G., Callaerts, P. and Gehring, W. J. (1995), Induction of ectopic eyes by targeted expression of the Eyeless gene in Drosophila, *Science*, 267: 1788–92.

Hepper, P. (2015), Behavior during the prenatal period: adaptive for development and survival, *Child Development Perspectives*, 9(1): 38–43; doi: 10.1111/cdep.12104

Hepper, P. G. (1988), Fetal 'soap' addiction, *The Lancet*, 11 June: 1347–8

Katz, L. C. and Shatz, C. J. (1996), Synaptic activity and the construction of cortical circuits, *Science*, 274: 1133–8

Lagercrantz, H. and Changeux, J.-P. (2009), The emergence of human consciousness: from fetal to neonatal life, *Pediatric Research*, 65(3): 255–60

Lecanuet, J.-P. and Schaal, B. (1996), Fetal sensory competencies, *European Journal of Obstetrics & Gynecology and Reproductive Biology*, 68 (Supp. C): 1–23; doi: 10.1016/03012115(96)02509-2

Mennella, J. A., Jagnow, C. P. and Beauchamp, G. K. (2001), Prenatal and postnatal flavor learning by human infants, *Pediatrics*, 107(6): E88

Quiring, R., Walldorf, U., Kloter, U. and Gehring, W. J. (1994), Homology of the Eyeless gene of Drosophila to the Small Eye gene in mice and Aniridia in humans, *Science*, 265: 785–9

Rosner, B. S. and Doherty, N. E. (1979), The response of neonates to intra-uterine sounds, *Developmental Medicine & Child Neurology*, 21(6): 723–9; doi: 10.1111/j.1469-8749.1979.tb01693.x

Schaal, B., Marlier, L. and Soussignan, R. (2000), Human foetuses learn odours from their pregnant mother's diet, *Chemical Senses*, 25(6): 729–37; doi: 10.1093/chemse/25.6.729

Webb, A. R., Heller, H. T., Benson, C. B. and Lahav, A. (2015), Mother's voice and heartbeat sounds elicit auditory plasticity in the human brain before full gestation, *Proceedings of the National Academy of Sciences*, 112(10): 3152–7; doi: 10.1073/pnas.1414924112

A Hairy Past

Bramble, D. M. and Lieberman, D. E. (2004), Endurance running and the evolution of Homo, *Nature*, 432: 345–52; doi: 10.1038/nature03052

Jablonski, N. G. (2010), The naked truth, *Scientific American*, 302(2): 42; doi: 10.1038/scientificamerican0210-42

Lieberman, D. E. and Bramble, D. M. (2007), The evolution of marathon running: capabilities in humans, *Sports Medicine*, 37(4–5): 288–90

Pagel, M. and Bodmer, W. (2003), A naked ape would have fewer parasites, *Proceedings of the Royal Society of London*, Series B: *Biological Sciences*, 270 (Supp. 1), S117

Powell, A. (2007), Humans: hot, sweaty, natural-born runners, *Harvard Gazette*, 19 April; https://news.harvard.edu/gazette/story/2007/04/humans-hot-sweaty-natural-born-runners/

From Water to Air

Bodmer, R. (1993), The gene Tinman is required for specification of the heart and visceral muscles in Drosophila, *Development*, 118(3): 719–29

Deglincerti, A., Croft, G. F., Pietila, L. N., Zernicka-Goetz, M., Siggia, E. D. and Brivanlou, A. H. (2016), Self-organization of the in vitro attached human embryo, *Nature*, 533: 251–4; doi: 10.1038/nature17948

Graven, S. N. and Browne, J. V. (2008), Sleep and brain development: the critical role of sleep in fetal and early neonatal brain development, *Newborn and Infant Nursing Reviews*, 8(4): 173–9; doi: 10.1053/j.nainr.2008.10.008

Li, W., Ma, L., Yang, G. and Gan, W. B. (2017), REM sleep selectively prunes and maintains new synapses in development and learning, *Nature Neuroscience*, 20(3): 427–37; doi: 10.1038/nn.4479

Louie, K. and Wilson, M. A. (2001), Temporally structured replay of awake hippocampal ensemble activity during rapid eye movement sleep, *Neuron*, 29(1): 145–56

Myrhaug, H. T., Brurberg, K. G., Hov, L., Håvelsrud, K. and Reinar, L. M. (2017), Prognose for og oppfølging av ekstremt premature barn: systematisk oversikt, Folkehelseinstituttet, Forskningsoversikt 01 2017. ISBN (electronic): 978-82-8082-799-9 Available at på www.fhi.no

Partridge, E. A., Davey, M. G., Hornick, M. A., McGovern, P. E., Mejaddam, A. Y., Vrecenak, J. D. and Flake, A. W. (2017), An extra-uterine system to physiologically support the extreme premature lamb, *Nature Communications*, 8: 15112; doi: 10.1038/ncomms15112

Schott, J. J., Benson, D. W., Basson, C. T., Pease, W., Silberbach, G. M., Moak, J. P. and Seidman, J. G. (1998), Congenital heart disease caused by mutations in the transcription factor NKX2-5, *Science*, 281: 108–11

Shahbazi, M. N., Jedrusik, A., Vuoristo, S., Recher, G., Hupalowska, A., Bolton, V. and Zernicka-Goetz, M. (2016), Self-organization of the human embryo in the absence of maternal tissues, *Nature Cell Biology*, 18(6): 700–708; doi: 10.1038/ncb3347

Shank, S. S. and Margoliash, D. (2009), Sleep and sensorimotor integration during early vocal learning in a songbird, *Nature*, 458: 73–7

The End — or The Beginning

BBC Earth (2014), Amazing birth of a baby kangaroo, 1 October; www.bbc.com/earth/story/20141001-newborn-baby-kangaroo/

Frank, L. G., Weldele, M. L. and Glickman, S. E. (1995), Masculinization costs in hyenas, *Nature*, 377: 584–5; doi: 10.1038/377584b0

Gao, L., Rabbitt, E. H., Condon, J. C., Renthal, N. E., Johnston, J. M., Mitsche, M. A. and Mendelson, C. R. (2015), Steroid receptor coactivators 1 and 2 mediate fetal-to-maternal signaling that initiates parturition, *Journal of Clinical Investigation*, 125(7): 2808–24; doi: 10.1172/JCI78544

Kota, S. K., Gayatri, K., Jammula, S., Kota, S. K., Krishna, S. V. S., Meher, L. K. and Modi, K. D. (2013), Endocrinology of parturition,

Indian Journal of Endocrinology and Metabolism, 17(1): 50–59; doi: 10.4103/2230-8210.107841

Lagercrantz, H. (2016), The good stress of being born, *Acta Paediatrica*, 105(12): 1413–16; doi: 10.1111/apa.13615

Lagercrantz, H. and Slotkin, T. (1986), The 'stress' of being born, *Scientific American*, 254(4): 100

Menon, R., Bonney, E. A., Condon, J., Mesiano, S. and Taylor, R. N. (2016), Novel concepts on pregnancy clocks and alarms: redundancy and synergy in human parturition, *Human Reproduction Update*, 22(5): 535–60; doi: 10.1093/humupd/dmw022

Nathanielsz, P. W. and Granrud, L. (1996), *Livet før fødselen*, Oslo: Pax

Trevathan, W. (2015), Primate pelvic anatomy and implications for birth, *Philosophical Transactions of the Royal Society*, Series B: Biological Sciences, 370(1663); doi: 10.1098/rstb.2014.0065